50 SIMPLE THINGS® YOU CAN DO TO SAVE THE EARTH

John Javna,
Sophie Javna, and Jesse Javna

50 Simple Things® is a Registered Trademark

HYPERION
NEW YORK

ISBN: 978-1-4013-2299-1

Hyperion Books are available for
special promotions, premiums, or corporate training.
For details contact Michael Rentas,
Proprietary Markets, Hyperion,
77 West 66th Street, 12th floor, New York, New York 10023,
or call 212-456-0133.

FIRST EDITION

10 9 8 7 6 5 4 3 2 1

We've provided a great deal of information about
practices, products, and different organizations in our book.
In most cases, we've relied on advice, recommendations, and
research by others whose judgments we consider accurate
and free from bias. However, we can't and don't guarantee
the results. This book offers you a start. The responsibility
for using it ultimately rests with you.

This book is printed on 100%
postconsumer recycled paper.

For more information, contact
www.50simplethings.com

THANK YOU

We sincerely thank the people whose
advice and assistance made this book possible.

Special thanks to the environmental groups that became our
partners in this book, and the people at those organizations who
worked closely with us to make this incredibly ambitious project
a reality. We'd love to list all 120 of you, but there simply isn't
enough room. So we just want you to know how much we
appreciate you, and the work you did on behalf of this book.
Thank you!

...And hats off to the rest of the crew:

Sharon Javna
Julie Bennett
Judy Plapinger
Will Balliett
Ken Wells
Cathy Hemming
Angela Kern
Claudia Bauer
Samantha Moss
Bob Kuenzel
Adam Siegel
Peter Miller
Susan Fassberg
Lorna Garano
Altemus Design
Joshua Redel
Kevin Davidsohn
Abe Genack
Donald R. Morrison
Brian Ayliss
Clare Butterfield
Wendy Millstine
Ben Clausen

Terry Griffin
Melissa Kirk
Eli Brown
Pat McFarland
Emily Bennett Beck
Joel Makower
Catharine Sutker
Jennifer Massey
Clint Willis
Original Design Solutions
Jennifer Strange
Brian Boone
Lou Brunette
Rachael Durfee
Julia Papps
Rhys Rounds
Alexis Soulios
Eric Stahlman
Brian Freeman
Thom Little
The Rogue Valley
Roasting Company
Manny the Dog

CONTENTS

A SIMPLE STORY: JOHN

I f you were a fan of the original *50 Simple Things You Can Do to Save the Earth* in the 1990s, you may be in for a surprise. This isn't an updated version of the original—it's an entirely new book.

Why did we change it? After all, it was one of the bestselling environmental books ever printed—more than 5 million copies were sold in 23 languages.

To tell the truth, it's because ultimately the book's approach didn't work.

I should say at the outset that I'm proud of the original *50 Simple Things*, and have always been honored that so many people found inspiration in it. But I've also been frustrated by its limitations—because while "simple things" like installing low-flow showerheads and taking cloth bags to a store can be a little piece of the solution to our environmental problems, they're not *the* solution. Eco-tips alone can never have a significant impact on "saving the earth." They're baby steps—and if they don't lead to something bigger, then we're in a world of trouble. Literally.

In retrospect, I can see that *50 Simple Things* didn't really educate people about the nature and extent of the enviromental problems themselves. As a result, many people believed that if they took a cloth bag to the supermarket instead of asking for a paper one, they were actually solving the problem of deforestation. Or if they snipped six-pack rings, they were preserving ocean fisheries. It created a false sense of complacency that these problems were being solved . . . when they weren't.

When I realized this in 1995, I decided to take the book out of print rather than update it. It hasn't been available for 12 years, but its influence lives on. For more than a decade, the public has been fed a steady diet of recycled *50 Simple Things*–style eco-tips, telling us how to use less paper, save energy at home, travel smarter, and so on.

As I've said, these suggestions have real value—in fact, you'll find some of them in this edition too—and people who practice them deserve credit for their efforts. But let's be realistic: After 17 years, where have the eco-tips taken us? They haven't taken mercury out of the air, or brought back songbird habitats. They haven't stopped polluted runoff from flowing into our waterways or coal companies from blowing the tops off mountains.

More important, while we've been focused on changing a few personal habits, the world's life-support system has shown increasing signs of collapse. Chances are, you're feeling overwhelmed by these developments. I feel it too, and this book is an effort to do something about it.

A LITTLE BACKGROUND

Many Americans are reluctant to become more involved with environmental issues because they're not "experts." Or they feel they're just not familiar enough with the facts to take action. I can tell you from experience that this doesn't have to be an impediment. I didn't write the original 50 Simple Things You Can Do to Save the Earth because I was an experienced environmentalist. Quite the contrary—I knew practically nothing about the environment except what I read in the newspapers. In the early winter of 1989, all I could tell was that everything seemed to be falling apart. There were front-page stories about acid rain, the ozone hole, global warming, and a host of other ecological problems that seemed to spell doom.

I became obsessed wondering if there was anything I could do about it. This was before the Internet, and there was no easy access to the collective knowledge of the culture. Instead, I went looking for a book that might guide me. I couldn't find one. The best I could find was a few lists put together by random groups, with suggestions for things like "Dance–it makes the earth a better place." It was a good idea, but I wanted slightly more practical advice.

I was almost 40, and was looking for a way to make my mark in the world. So I decided to write the environmental book I'd been looking for—a guide to the things any individual could do to save the Earth. I'd been an author for about 10 years by then, but none of my books were what you'd call serious. My bestsellers were The TV Theme Song Singalong Songbook and Uncle John's Bathroom

Reader. You can imagine how my idea was received by publishers. The nicest response I got was "Huh?"

So I did what many determined authors do: I borrowed money and published it myself. I was going to call it *100 Simple Things You Can Do to Save the Earth*, but I chickened out. What if there weren't 100 things? I cut it to 50.

THE FACTS OF LIFE

The book was released without fanfare in the fall of 1989, and I wasn't at all prepared for what happened next. People—apparently hungry for a hopeful, concrete approach to environmental problems—started snapping it up all over the country. Seemingly out of nowhere, millons of copies were sold, and within a few months the book had appeared in all the major media, including the *New York Times* bestseller list—where it debuted at #1.

It was an exciting time for me—particularly because the green wave sweeping America in the spring of 1990 convinced me that we were going to change the world. I thought the 20th anniversary of Earth Day—in which millions of people participated—and the success of *50 Simple Things* were part of a grassroots revolution: Now that we knew how important it was, we'd all rush out to buy recycled products and faucet aerators, and transform the economy into a green machine. But we didn't. Earth Day came and went, and the media shifted its focus. Gradually, over the next few years, public enthusiasm waned—depressed by clever greenwash campaigns staged by plastics, chemical, and oil companies, that confused people just enough to slow down the movement's momentum.

By the time a book called *Simple Things Won't Save the Earth* came out in 1994, I was inclined to agree with the author. I'm embarrassed to admit it now, but I'd become pretty cycnical. I had a case of "green fatigue"—I'd had one eco-tip too many. The cynicsm didn't come from not caring, it came from feeling that no matter how many tips I tried, the problems were so big that I could never make a dent in them. What did it matter if I recycled paper, if the ancient forests were still being chopped down? Who cared if I celebrated "no car day" when 80% of the cars on the road had one person in them? There was mercury in the air, and tons of waste was being dumped in the ocean daily. Every time I looked in the garbage and saw a pile of aluminum cans, I felt like giving up. And then, like a lot of disillusioned and overwhelmed Americans, I *did* give up. I moved to rural Oregon and focused on raising my family.

TWELVE YEARS LATER

In 2006, my 13-year-old daughter, Sophie, started to become environmentally aware. She began asking why we didn't compost anymore . . . and why I didn't bring cloth bags to the supermarket. One day, I started to tell her why it didn't matter—why all the well-meaning recycling in the universe wouldn't stop global warming. But I stopped in mid-sentence. It was weird—I found myself staring, literally, into the eyes of the next generation, the person I had written my book for years before she'd been born. It dawned on me that I couldn't afford to be cynical—I had to keep trying to make the world better—because I love my son and daughter, and because I love this planet.

That epiphany was the genesis of the book you're holding in your hands. It's a father's effort to reclaim the Earth for his children, and yours.

INVENTING A NEW BOOK

With my newly discovered enthusiasm, I decided that the most effective thing I could do as an activist was to bring back *50 Simple Things*. But the only part of the original I wanted to retain was its simplicity, its user-friendliness. The rest had to be reinvented. My co-authors and I wrote with these four guiding principles in mind:

1) The actions in this book need to be framed in terms of issues. No random, piecemeal actions. There are problems to solve; we need to know what our goals are, and how to get there.

2) The individual efforts that really make a difference on behalf of the Earth are sustained, committed ones. Random acts of environmental kindness are fine, but we're out to turn things around. That means finding ways to make a long-term commitment comfortably—without turning ourselves into green monks.

3) Individual action needs to be combined with community action. People are strongest and most effective when we join forces with our neighbors, especially on big issues—and this is the biggest issue any of us will ever face.

And finally,

4) We need to help focus readers' efforts. Environmental problems are so overwhelming, and there's so much to do, that it's hard to know where to start. It's our job to provide the entry point.

The question was, could a "tips" book like *50 Simple Things* really accomplish this? My co-writers and I put a lot of thought into it, and finally came up with something no one's ever tried before: We turned our book into an interactive partnership between individual readers, environmental organizations, and us. The new *50 Simple Things* is not just a book, but a doorway into a community of experts and grassroots activists who can help you accomplish more than you could possibly hope to do on your own. And to support it, we've created a vibrant, new online community. Our new Web site, ***www.50simplethings.com***, is designed to be an ongoing source of information about what you can do—an opportunity to learn, ask questions, and share what you've learned. We've done our best to make sure this book lives up to it goals. Now it's up to you.

THE CONNECTION

One day, after we'd started working on this book, Sophie asked if the issues we'd included were big enough. "I know if I was going to pick one thing to work on," she said, "it would be the biggest, most important thing possible." I thought for a minute, and said to her what I'd like to share with you now: The "big things" we have to tackle, like global warming or habitat loss, aren't "things." They're effects—the results of a lot of little, destructive actions. The only way to deal with them is to undo them one at a time, the way we created them.

And truthfully, it doesn't matter which issues we choose to work on—big or small—because they're all connected. If, for example, we reduce greenhouse gas emissions from power plants, we also help clean waterways; if we help clean waterways, we improve wildlife habitats; to improve wildlife habitats, we need to set aside wilderness; and in setting aside wilderness, we protect trees. When we have more trees, we clean the air. And better air means less climate change. It's the circle of life, and it holds us all. Just pick a spot and jump in.

SOPHIE'S VOICE

Sophie Javna is 14 years old. She's a high-school freshman.

One of the most engaging parts of childhood is dreaming about the future. We all do it: What am I going to be when I grow up? Will I be rich and famous? Will I get married and have kids?

Now that I'm a teenager it's especially fun, because I understand all the possibilities that stretch out in front of me. I'm starting to have a real idea of what life might be like as an adult. Maybe I'll even pursue my dream of becoming a professional singer, have a record deal and a band. Whatever happens, I know it will be exciting.

My perfect future sounds great...until I come back to Earth and start thinking seriously about what the world will be like 10 or 20 years from now: Will there be endless amounts of pollution...dead zones in the ocean...no polar bears, elephants, gorillas, cheetahs... our beautiful rainforests completely gone? I don't know—maybe the future isn't really all it's cracked up to be.

To me, this environmental problem seems big enough for people all over America to rally together and ask, "How can we fix this?" And if you're reading this, you probably think it is, too. But an amazing number of people still seem to have no idea how incredibly important it is—or, even worse, they know, and still go on with their lives pretending they don't. Every day, I see signs of those people; their Hummers lumbering down the street, their bottles and cans filling the trash. Then I read about politicians who just don't seem to understand how urgent it is to do something about global warming, about clean water, and even about safe food.

Whenever I see these things, I can't help but feel a deep hurt. Because all their actions are really saying to me is, "I don't care. I don't care about the world, and I don't care what your life—or any child's life—will be like when *I* am no longer alive."

Children have the ability—even the desire—to believe that anything can happen. I'm not too old for that, even at 14. I honestly believe we can change our future for the better, and I know that if we really try, we can all come together and figure this thing out.

There is no doubt in my mind that every single person is capable of doing their part to save our planet, if they just get a little push in the right direction. So even though things may not be going the way we'd like them to with the environment, I'm not going to start thinking negatively. That is not what this book is about. It's about being positive, saying we can join forces and work towards the same goal, and—because anything is possible—we will solve this problem.

By working to save the environment, we're protecting my generation's hope for a better future and a better life. And, just in advance, I want to thank you for that.

—Sophie Javna,
January 2008

A MATTER OF SECURITY

Jesse Javna is 17 years old. He's a high-school senior.

Every generation has had its own vision of the future. My parents grew up watching TV shows about spacemen with flying cars and walkie-talkies, and robots that made life easy. People who grew up with *Star Wars* were energized by the belief that good would inevitably triumph over evil. In both cases, the future ultimately seemed like a time of security and optimism.

Now that I'm almost officially a young adult, I've begun to think about my own generation and where we fit in the world. What is our vision for the future? Do we even have one?

I have to confess that in recent years, our future has begun to look pretty grim—not because members of my generation don't look forward to experiencing life, but because the problems we face are so daunting. I'm part of the 9-11 generation; we are the first American kids since Pearl Harbor to feel that maybe the United States isn't as strong or as secure as we'd imagined.

We are also at the heart of the Global Warming Generation. No other generation before us has had—from our first moments of social consciousness—to seriously confront the possibility that the world might not exist in its current state because of human impact on the environment.

Sometimes it feels like the environmental problems we face are impossible to overcome. It seems like we're helpless to fix the damage that has already been done, and that we're not even taking small steps to prevent further harm to our home planet.

It's no wonder that as we begin to enter the world as adults, many people in my generation are approaching the future in fear. But where will that get us? If we are afraid of the future, if we choose to ignore the challenges we face, the problems will only get worse. We can't be afraid of taking charge of the world—we have to lead with optimism. The best way to do that is to act now. In the long run, our only real security—national, global, and personal—will come from knowing that we're doing everything we can to change things for the better.

You may not know where to start, and you're not alone. A lot of people don't know where to begin when it comes to making a positive change. But that's exactly what this book is designed for—to give people like you and me, who want to do something *now* to protect our world, a way to find what matters to them and to make a difference.

We don't have to wait for older generations to make the first move. It's our turn to take charge and secure our own future. Earth is our only home, and together, we have the means to save it.

—*Jesse Javna,*
February 2008

HOW TO USE
THIS BOOK

*Here are a few things we'd like you
to know before you get started.*

BACKGROUND. There are plenty of environmental "how-to" books around. This one is different: It's not a list of things to do in your home or office, and we don't suggest that you do as many as possible. This book is made up of 50 separate environmental issues...and we encourage you to pick just one.

Yes, just one.

You can do more if you want, of course. But let's be realistic: One of the main reasons most people don't get involved with the big issues confronting us—and nothing is bigger than protecting our life-support system—is that the issues seem too overwhelming. We think, "Where do I start?" "Can I really make a difference?" "There's too much to do." And, finally, "Why bother?"

That's what we have to overcome.

Years ago, we helped to popularize the notion that the things one person can do on their own can have an impact on the environment. Now we want to add a new piece to that idea: Picking one thing to work on, and making a sustained, committed effort on that issue. It's the most satisfying and effective way for you to help take care of the planet—maybe the only way.

How committed do you have to be? That's up to you. Experiment. Find your level of comfort, stick with it, and celebrate it. Don't let anything make you feel guilty about "not doing enough."

Environmental action isn't something to avoid. It's way to confirm our love of family, community, the Earth, and life itself. This book is designed to make it easy for you to find something you love to do, and to keep doing it.

ABOUT THE 50 THINGS
• We've included a wide variety of issues and activites—from working in the wilderness to working in city hall; from meeting

with your congressperson to talk about legislation, to meeting with your neighbors to discuss planting trees. There's something for policy wonks, and something for naturalists. But, of course, these 50 issues aren't the only ones that matter—there are plenty more that could have been picked.

• So how did we pick them? Actually, we didn't. We contacted 50 of America's most respected environmental groups and asked each of them to pick one issue they wanted to share with our readers. For some groups it was easy; they focus primarily on one area. For others, like the Natural Resources Defense Council or the Sierra Club, it was harder because they work on so many different issues. But they all got into the spirit of the project, and this book is the result.

• The 50 issues are numbered, but there's no special order to them. They're not a "Top 50"—none is more important than any other.

ABOUT THE PARTNERS

• The groups we chose to work with are not necessarily the biggest, or oldest, or most influential. We've included a cross-secton of the American environmental movement: Some are well established, like the National Audubon Society. Some are less well known, like Seacology and Eco-Cycle. Some are even brand-new, like 1Sky. Large or small, they're all among the most respected organizations in their fields.

• On each of the issues, we worked closely with experts from the group that chose it to develop a presentation and an action plan.

• Although we selected these 50 organizations, there are many more that could just as easily have been included. You'll find Web links to many of them at 50simplethings.com

ABOUT THE PAGES

• Most of the book is self-evident, but a few things are worth mentioning: "Start Simple" is a way to test the waters, to get familiar with the issue and see whether it's something that interests you. It's a way to build a bridge between the thought, "Hmm, that sounds interesting" and actually doing something about it.

• "Steps for Success" is an overview of some of the things you can do regarding each issue. The first actions are generally the easier tasks; they work up to the harder ones.

• There's no time limit for doing the steps. It may take a year to do Step 1 and another for Step 2. It doesn't matter. What matters is that you keep at it, not speed through it. Finding ways to make the issue part of your life is what commitment is all about.

• As you'll see, almost every action ends with a Web address. That's because this book is only the beginning—there's more to learn, and more to do at our partners' Web sites, and at our own. We've built the Net into the book as an essential resource.

ABOUT THE INTERNET

• A 50 Simple Things Web site has been created as a companion to the book (*50simplethings.com*). At the top of the right-hand page in each chapter, there's a Web address such as *50simplethings.com/streams*. Each will take you to a page specifically related to that chapter: links, downloadable guides, discussions and blogs, feature articles, and so on.

• On our Web site, you'll find a copy of the original *50 Simple Things You Can Do to Save the Earth*. You can also find footnotes, a blog, links, and general information on environmental action, plus lots of other features we'll be developing.

• Visit our partner organizations' Web sites. They generally offer extensive information about the subject they've sponsored, and in many cases have added information specifically for you.

• We've done everything we possibly could to supply accurate links and Web addresses. If you find that one doesn't work, please notify us immediately. We'll post any corrections necessary. **Note:** We've listed Web sites without the traditonal "www." The w's aren't usually necessary, but if a site doesn't work, try adding the w's to see if it makes a difference.

ONE LAST THING

Our strong advice: *Don't read this book cover to cover.* Dip into it, learn about the issues one at a time. Visit the Web sites referenced, and explore the topics in depth. If you take your time, you'll find what you're looking for. Good luck...and have fun!

50
SIMPLE

THINGS

1. BRING BACK THE ELECTRIC CAR

Estimated fuel cost for an all-electric car to go 40 miles: about 85¢
Fuel cost for a gas-powered car to go the same distance: $5.60

BACKGROUND. Wouldn't it be great if your family car was electric, and you could just plug it in instead of taking it to a gas station? Well, you can't go out and buy one from your local car dealer right now, but there are dozens of companies working on an improved electric battery...and major carmakers say they're developing "plug-in hybrids"—cars that run on electricity, but switch to gasoline when the battery's charge is used up.

What these pioneers need right now is public support. It's important for automakers to know that we really want electric cars...and that we'll buy them as soon as they're available.

DID YOU KNOW

• The first cars in America were electric. Until Henry Ford made the combustion engine more affordable, electric cars were just as popular. Since then, auto companies have ignored electricity in favor of cheap oil.

• That may be why people don't believe electric-car technology exists. But it's here, and it's been tested. GM, Honda, and Toyota all developed electric cars in the late 1990s, and the lucky people who drove them loved them. Now GM and Toyota say they'll have plug-in hybrids on the market by 2010.

• The benefits of plug-in electric cars are impressive: They're cleaner, cheaper, and fueled by domestic sources of energy. Even with today's power plants, a plug-in car with an "all-electric range" of 40 miles can put out 50% less greenhouse gases than a conventional car. If that electricity is generated by renewables, the number skyrockets to almost 80% less than a regular car.

• Electricity is cheap—you'll pay less than $1 for the energy equivalent of a gallon of gas. And many utilities plan to offer special low rates for people willing to charge car batteries at night.

In 2008, Israel announced its support of the world's first electric car network. Target date: 2011

WHAT YOU CAN DO

Your Partner: Friends of the Earth (FOE) is the U.S. voice of an international network of grassroots groups in 70 countries. Get acquainted at *PlugInNow.org*

Your Goal: Use your influence as a car buyer to convince carmakers to invest in electric technology. Let them know people want electric cars. "The key to success is the automakers," says FOE's Kate Horner. "They'll only get behind these cars if they feel there's a market for them. Once they see how many of us are waiting for plug-ins, the auto industry will be changed forever."

START SIMPLE

Movie night. In 1990, California passed a law requiring "zero-emissions vehicles." Automakers were forced to develop all-electric cars, which they leased to the public as "demonstration vehicles." You can't lease one anymore, because GM crushed its electric cars—literally—despite owners' pleas to keep them. Rent the film *Who Killed the Electric Car?* for this fascinating story. Better yet, buy a copy at *PlugInNow.org*, watch it, and pass it on to friends.

STEPS FOR SUCCESS

Step 1. Fight the myths about electric cars. Learn more about them; become a salesperson for the concept. Read "Electric Cars 101" at *PlugInNow.org*

Step 2. Join FOE's Electric Car Club. Sign up at FOE's Web site. They'll send you updates on the latest electric vehicle developments and tell you how you can help.

Step 3. Talk to automakers, especially GM and Toyota—the two most likely to build plug-in hybrids. Sign FOE's online petition saying you'd consider buying a plug-in hybrid and mail letters to GM and Toyota telling them you're interested in buying one. (Their e-mail and snail-mail addresses are at FOE's site.)

Step 4. Talk to your legislators. Make sure the government provides the same benefits to plug-ins and all-electric vehicles that they have with hybrids, like tax credits ($2,000–$3,000) and research grants. Get the facts at *PlugInNow.org*

For more resources: *50simplethings.com/electriccar*

In a modern electric car, 98% of the battery is recyclable...and it doesn't contain lead.

2. GO ORGANIC

By planting only 10% more organic food, we can transform over 25,000 square miles of depleted soil into rich, highly productive cropland.

BACKGROUND. It wasn't long ago that most Americans thought of organic food as something only "health nuts" ate. But today, about one in four U.S. families buy organic products every week. Organics is the fastest-growing segment in the food market.

This is great news for people who care about the environment, because organic farming is the essence of sustainable agriculture. It builds soil, provides habitat, and eliminates pesticide runoff. But the movement is just beginning to get firmly rooted. We need to help make it grow.

DID YOU KNOW

• "Organic" means something is grown without "toxic and persistent pesticides and fertilizers." It also refers to a method of agriculture that uses natural systems to maintain and replenish topsoil.

• Organically produced foods are also grown without antibiotics, synthetic hormones, genetic engineering (GE), sewage sludge, or irradiation. (Yes, believe it or not, it's common in America to use *sewage sludge* on crops.)

• In 2006, scientists reviewed 76 different studies that compared conventional and organic farming. They created 20 categories to analyze. Organic and conventional were about equal in 8; organic was better in the other 12. That included organic farms having an average of 30% more biodiversity, 24% less soil erosion, and, of course, 100% less pesticide runoff.

• New studies confirm other benefits. In 2007, a $24 million, four-year experiment conducted for the European Union showed that organically grown produce has considerably more antioxidants, and higher levels of vitamins. "The health benefits were so striking," one critic explains, "that moving to organic food was the equivalent of eating an extra portion of fruit and vegetables every day."

The USDA Organic seal on food officially debuted on October 21, 2002.

WHAT YOU CAN DO

Your Partner: The National Campaign for Sustainable Agriculture (NCSA). They help individuals and organizations build "healthy, environmentally sound, profitable, and humane" rural communities and agricultural systems. Get acquainted at *sustainableagriculture.net*

Your Goal: To promote and support organic, sustainable agriculture one step at a time.

START SIMPLE

• *Buy organic!* Start small; pick one organic item to replace a product you usually buy. Start with soft-skin fruits, vegetables, or nuts. Tests show they absorb the most pesticides.

• *The Shopper's Guide* is a wallet-size card that lists the most-contaminated fruits and vegetables, as well as the most "consistently clean" items. Download it at *foodnews.org*

STEPS FOR SUCCESS

Step 1. Use your consumer power. Build the market for organic food and products by buying them for your family. Start with produce, then add staples like milk and bread. Think of it as an investment in a cleaner environment and a healthier food system.

Step 2. Lobby locally. Grassroots tactics will help build the movement: Talk to store managers; bring organic brownies to a meeting or school event; plant a small organic garden; encourage your favorite restaurant to use more local, organic food.

Step 3. Support organic farming. Current government programs subsidize only the biggest farms. Encourage lawmakers to create laws that support organic farming; add your voice and efforts to local, state, and national groups working to ensure a healthy food system. For more info: *sustainableagriculture.net/organic.php*

Step 4. Protect organics. The National Organics Program's (NOP) "USDA Organic" label guarantees that the product meets high standards. Be a watchdog: Get on the *Organic Action Alert* list run by NCSA; help protect and strengthen the organic label and ensure proper enforcement of NOP standards: *sustainableagriculture.net*

For more resources: *50simplethings.com/organics*

According to the U.K. Soil Association, 50–93% of pesticide residue remains on broccoli after washing.

3. HERE COMES THE SUN

The total amount of energy that comes to Earth from the sun each year is enough to provide more than 10,000 times the energy we use globally.

BACKGROUND. Do you think solar power is just a dream for the future? Then you've got a pleasant surprise coming: Reliable solar technology is available *right now*.

In fact, at a rate of over 30% a year, solar is one of the fastest growing energy sources in the world…and for good reason: It's free, it's renewable, it's plentiful, and it doesn't pollute. What more could we want?

DID YOU KNOW

• In 2006, the solar power industry grossed over $16 billion. But most of that wasn't spent in America—it was spent mostly in Germany (about 50%) and Japan. The reason: Both countries are subsidizing solar to drive down the cost and jump-start the market.

• Germany's solar industry, the biggest in the world, generates over 10,000 jobs in production, distribution, and installations. And they're fighting global warming at the same time: Using solar power to supply a million homes with energy, for example, reduces CO_2 emissions by about 4.3 million tons per year—the equivalent of taking 850,000 cars off the road.

• *Our* federal government, on the other hand, is doing very little to promote or support solar energy—even though the U.S. has always offered incentives to emerging energy industries. In fact, we *still* give billions to Big Oil (one estimate: $15–35 billion a year in subsidies!) and nuclear power companies (tax credits, loan guarantees, and insurance through the Energy Policy Act of 2005).

• Without government support, the U.S. will lose green-collar jobs to other nations that are investing in solar as a national priority.

• Experts predict that with the right government incentives, the cost of solar could fall rapidly over the next few years, and within a decade, solar could be about as cheap as fossil-fuel-based energy. "At that point," says General Electric's chief engineer, "you can expect pretty much unbounded growth" of solar power.

Today's most successful commercial application of solar energy in the U.S.? Swimming pool heaters.

WHAT YOU CAN DO

Your Partner: The American Solar Energy Society (ASES), a national organization dedicated to advancing the use of solar energy and bringing solar supporters together to create a sustainable energy economy. Get acquainted: *ases.org*

Your Goal: Be a "solar advocate." Work with ASES to strengthen education and outreach efforts needed to bring solar into the mainstream. Spread the word that this is the science that can save us.

START SIMPLE

• **Prove it to yourself.** Want proof that solar technology is viable? Go online and Google "solar products"; see what's available.

• **Go on a solar tour.** Communities all over the country offer tours of homes and businesses that have converted to solar. It's really inspiring. Most tours take place over the first weekend in October. Find the list of tours all over the country at *NationalSolarTour.org*

STEPS FOR SUCCESS

Step 1. Bring solar to your family and friends. "By learning and talking about solar," says ASES president Brad Collins, "you'll already be helping to make it grow, because the biggest drawback (besides cost) is that people don't understand how *real* it is." ASES will supply all the resources you need to become solar-energy literate. Start with ASES's Web site. For more info and a variety of links, go to *50simplethings.com/solar*

Step 2. Be an advocate. Government officials have to hear, over and over again, that *the people* want solar power, and that we want the government to invest in it *now*. Help elected officials stand up to Big Energy lobbyists who insist everything is just fine the way it is. Sign up for "solar action alerts" that will inform you whenever there's a legislative showdown you can be part of. Go to *ases.org/about.htm* and click "Get Email Updates."

Step 3. Bring solar to your community. While the cost of conventional types of energy go up, the cost of solar keeps coming down. Find out what solar rebates and incentives are available where you live; join forces with other pro-solar volunteers to push for *new* incentives; explore solar financing options. And if you can install solar at home, do it! ASES will help: *ases.org*

e World Bank estimates that the global market for solar electricity will reach $4 trillion in about 30 years.

4. SUPPORT SUSTAINABLE FORESTRY

Forests once covered at least 48% of the Earth's land surface. Half of them are already gone.

BACKGROUND. Old forests feel magical. It seems like they've always been there...and somehow they'll survive forever, no matter what we do.

But that's an illusion. About 95% of the ancient forests that existed in America 200 years ago are gone. As demand for forest products grows, the pressure to harvest more of America's natural forests keeps increasing. If we want them to survive, we have to make sure they're harvested sustainably.

DID YOU KNOW

• Forests are more than just trees; they're complex ecosystems—delicate balances of plants, animals, microorganisms, soil, and water. They support about two-thirds of all life on the planet.

• "If a forest is managed sustainably," says Richard Donovan of the Rainforest Alliance, "it can last forever. Even if it's logged, a sustainable forest will be healthy enough to support life."

• Case in point, according to the NRDC: "The Collins Almanor forest in California contained 1.5 billion board feet of standing timber when harvesting began in 1941. Sixty years and 2 billion board feet later, this sustainably managed forest still holds 1.5 billion board feet of standing timber, and supports great blue heron rookeries, black bears, rubber boas and bald eagles."

• Not all "sustainable" forests are equal. To some American lumber companies, a sustainable forest is one that keeps producing wood. So they have no problem cutting down old forests, destroying their ecosystems, and replacing them with fast-growing tree plantations.

• Are those two forests equivalent? Not even close. "A plantation forest is more like a field of corn than a natural forest," says Donovan. "Unless a plantation is designed very carefully to protect ecosystems, it will be a hostile environment to nearly every animal, bird and even insects. And it's been shown to have a negative

Deforestation accounts for one quarter of all greenhouse gas emissions globally.

impact on the water cycle because nonnative, fast-growing trees use high volumes of water. Pesticides are also commonly used."

• So although the total amount of trees in the U.S. is rising slightly, we're still losing forests. "And what we're not losing, we're degrading," Donovan explains. "They're taking out all the good trees and leaving the sickly ones, which degrades the seed base so only the weakest trees reproduce. It's a Darwinian nightmare."

WHAT YOU CAN DO

Your Partner: The Rainforest Alliance believes it's possible to create a sustainable society that protects the planet and provides sustainable livelihoods for its people. Visit *rainforest-alliance.org*

Your Goals: Find ways to protect America's forests and make sustainable forestry commercially viable.

START SIMPLE

Find 100% postconsumer recycled copier paper at a local office supply store. Chains like Staples or Office Depot normally carry it (call ahead to be sure), and it's probably more affordable than you think. Encourage others to do it, too. It's not just theoretical: By using paper that's been used before, you're saving trees and forests. For more information, go to *50simplethings.com/forests*

STEPS FOR SUCCESS

Step 1. Find out about sustainable forestry. In 1993, a group of businesses and environmental groups created the Forest Stewardship Council (FSC) to certify products from sustainably managed forests. The timber industry invented its own group, Sustainable Forestry Initiative (SFI), which condones practices like clear-cutting. Learn the difference: Go to *rainforest-alliance.org* and click on *forestry*. Other sites: *fsc.org, dontbuysfi.com, 50simplethings.com/forests*

Step 2. Support sustainable forestry. Buy wood and wood-related products (even maple syrup) certified by the Forest Stewardship Council (look for *FSC* on the product) and the Rainforest Alliance (look for a frog logo). By supporting local FSC businesses with your dollars, you're contributing to the long-term sustainability of working forests and the communities that depend on them. Check for the FSC logo on catalogs, too. For info, check out *fsc.org* or *rainforest-alliance.org/forestry*

Support forests: Sign up for legislative alerts at *sierraclub.org, nrdc.org,* or other forest protection groups.

5. SAVE THE CORAL REEFS

A recent study found that about 70% of the world's coral reefs are already either threatened or destroyed.

BACKGROUND. Once you've seen a coral reef, you'll never forget it: the intricate patterns, the vibrant colors. But coral reefs aren't just beautiful; they're one of the Earth's most valuable habitats. They cover only about 1% of the planet's surface, yet they're home to 25% of all known marine life.

Unfortunately, they are also the planet's most endangered ecosystem. About 25% of our coral reefs are already gone. Experts say that within the next century they could disappear altogether... unless we do something to save them.

DID YOU KNOW

• A coral reef is a huge underwater limestone deposit created by colonies of fragile, tubular-shaped creatures called *coral polyps*. Normally, polyps continuously build coral, and a reef grows and thrives. But when conditions become too adverse, the polyps die.

• Coral death threatens other marine life as well. Over 4,000 species of fish, such as parrot fish and angelfish, inhabit reefs. Jellyfish, lobsters, starfish, sea urchins, turtles, and countless other creatures call reefs home...and depend on them for food.

• Coral reefs have existed in their current form for about 50 million years...which makes their potential destruction even more alarming. The main threat: global warming. Increased carbon dioxide in the air is making the ocean not only too hot, but also too acidic. (The ocean absorbs CO_2, which turns into carbonic acid.)

• Scientists say that coral is resilient and may be able to adapt to new conditions—but only if we reduce man-made stresses such as pollution, erosion (silt smothers coral), development (pesticide runoff), and overfishing (dynamite, cyanide, and bleach are used).

• So far, researchers have examined only a small part of reefs' potential benefits. For example: Recently discovered compounds found in coral are being used in AZT, the anti-AIDS drug, and 50% of new cancer research focuses on marine organisms.

WHAT YOU CAN DO

Your Partner: Seacology, a group dedicated to preserving the ecosystems and cultures of islands throughout the globe. They've saved over 1.7 million acres of coral reefs since 1993. Get acquainted at *seacology.org*

Your Goal: Find ways to protect coral reefs by increasing awareness of the risks to reefs, and taking action to protect them—even if you live 1,000 miles from the ocean.

START SIMPLE

Movie night. Even if you can't get to the ocean in person, you can still experience the beauty of a coral reef. Rent or buy a copy of the inspiring movie *Coral Reef Adventures*. Share it with friends. For info on this and other films, go to *50simplethings.com/reefs*

STEPS FOR SUCCESS

Step 1. Protect your own reef. Seacology saves reef areas by trading for them. They provide something an island society needs (like a school) in exchange for a "no-take" marine preserve area that protects reef areas from fishing. Seacology will establish a *50 Simple Things* fund for readers of this book; 100% of your donations will go toward saving a specific reef. Seacology will keep you informed and send photos, so you'll see exactly how your money is spent. Get the details at *seacology.org/50simplethings*

Step 2. Inspire a child. Seacology recently began an Adopt-An-Island Classroom program. It's an easy, fun, and rewarding way for students and teachers to not only learn about coral reefs, but take an active role in saving them. You can introduce this program to your local schools and help create a whole new generation of coral reef advocates. Details: *seacology.org/education*. For other ways to help save coral, no matter where you live: *50simplethings.com/reefs*

Step 3. Fight for more Marine Preserves. Only .01% of the ocean is protected from fishing, dredging, and dumping. Join forces with Seacology and other coral conservation groups that are trying to get governments to create more marine parks and preserves. Find links to them at *50simplethings.com/reefs*

Gardeners' alert: Nitrogen-based fertilizers eventually drain into the ocean and damage coral reefs.

6. AN INVISIBLE THREAT

It takes only 1/70th of a teaspoon of mercury to pollute a 25-acre lake so badly that its fish will be unsafe to eat for a year.

B ACKGROUND. Take a look outside. You can't see mercury floating in the air, but it's there…and there's a chance it's affecting your health.

Mercury is a heavy metal that's spewed into the atmosphere by industrial polluters like coal-fired power plants, incinerators, and cement kilns. Rain and snowmelt wash it into our waterways, where it's absorbed by plants. It gradually works its way up the food chain, contaminating fish and poisoning larger animals like polar bears, and humans.

America has affordable technology that can eliminate most man-made mercury pollution…but not everyone's using it. You can help make sure they do.

DID YOU KNOW

• Research shows that it takes only *a milligram* of mercury to harm a person. Yet each year we allow coal-fired power plants to emit at least *48 tons* of it. Power plants are considered the single largest source of mercury emissions.

• Mercury is now so prevalent in our waterways that at least 40 states have issued advisories restricting fish consumption from their rivers, lakes, or coastal waters.

• Lab studies on birds show that mercury damages their central nervous systems and is toxic to their developing embryos. It also seems to weaken immunity, making wild birds more susceptible to disease.

• Mercury's effect on people is equally alarming. It targets our nervous systems and kidneys. It's particularly dangerous for fetuses; they can absorb mercury from their mothers and suffer irreversible mental damage. The risk is real: About 8% of 20–40-year-old women have enough mercury in their bodies to harm their unborn babies.

WHAT YOU CAN DO

Your Partner: Earthjustice, a nonprofit public-interest law firm that's been representing environmental groups since 1971, "because the Earth needs a good lawyer." Get acquainted at *earthjustice.org*

Your Goal: Force polluters to reduce or eliminate mercury emissions. "There are different kinds of environmental actions," says Earthjustice's Jim Pew, "and recycling a newspaper is *not* going to get mercury out of the air. We're talking about major industrial polluters. It's the government's responsibility to deal with them, and it's our job as citizens to make sure they do."

START SIMPLE

Power point. Is a coal-fired plant spewing mercury into the air every time you turn on a light? Most of us have no idea...which is one reason this kind of pollution continues. Earthjustice will help you find out where your electricity comes from and how it's generated. It's important for each of us to know: *earthjustice.org/location*

STEPS FOR SUCCESS

Step 1. Say it loud. If mercury's a problem in your area, let people know. Building awareness is a powerful weapon when you're trying to influence the government. Spread the word through Letters to the Editor, conversations with friends and neighbors, and blogs. Find suggestions at *earthjustice.org/mercury*

Step 2. Be a watchdog. Power plants periodically have to get their state "clean-air" permits renewed. When that time comes, your local environmental agency will post notices asking for public comments. Keep watch online: When you get the notice, rally people in your area to join you in commenting. It's your chance to be heard. Info on how to do it is at *earthjustice.org/mercury*

Step 3. Use your influence. We have technology that can clean up mercury emissions by 90%. Why aren't our government representatives *insisting* that it be installed? If the feds made power plants follow existing cleanup laws, mercury could largely be controlled within a few years. Get tips on what to say to elected officials, and how to reach them, at *earthjustice.org/lobbying*

For more resources: *50simplethings.com/mercury*

Each year, some 600,000 U.S. babies are at risk for brain damage from their mothers' exposure to mercury

7. VOTE FOR
THE EARTH

U.S. Rep. Richard Pombo (R-CA) ruthlessly fought to roll back environmental protections in Congress. In 2006, environmental groups ran a door-to-door campaign against him, informing voters of his positions. He was defeated by a "green" candidate.

BACKGROUND. If we want to protect the environment, we've got to have government support. On every level—local, state, national—it's elected officials who determine which regulations get enforced, which laws get priority, and which new industries or technologies get subsidized. We need to make sure those elected officials are on our side.

Candidates don't get elected by themselves. There's a team of people behind them working for their success—from campaign managers to voters. If politics interests you, pick candidates who'll fight for the health of our planet...and work like hell for them.

DID YOU KNOW

• Does it really matter who's in office? You bet. Our elected officials also appoint the people who enforce the rules...so if you voted for George Bush in 2004, you also voted for Philip A. Cooney—a lobbyist for the American Petroleum Institute who served as chief of staff for the Bush administration's White House Council on Environmental Quality (no kidding).

• If you voted for Sen. James Inhofe (R-OK), you put a man in office who believes global warming is the "greatest hoax ever perpetrated on the American people."

• If you voted for Rep. Don Young (R-AK), you put a man on the House Committee on Natural Resources who told a reporter, "When I see a tree, I see paper to blow your nose."

• State and local candidates can make as much of a difference as national candidates. They set priorities on issues like wetlands, recycling, and renewable energy. For example: Salt Lake City's climate program has already reduced local greenhouse gas emissions by 31% since 2001, well below the targets of the Kyoto Protocol.

Believe It or Not: Matthew Hogan, a lobbyist for big game hunters, was nominated...

WHAT YOU CAN DO

Your Partner: The League of Conservation Voters (LCV), the "political voice of the environment," has worked for pro-environment candidates and sound environmental policies since 1970. *Your Goal:* Work with LCV on their tough, effective campaigns to defeat anti-environment candidates, and support those leaders who stand up for a clean, healthy future for America.

START SIMPLE

Register to vote. Make sure every environmentalist you know is registered, too. Tips for a registration drive: *50simplethings.com/vote*

STEPS FOR SUCCESS

Step 1. Find out where candidates stand on the environment. Get LCV's National Environmental Scorecard. It will inform you about the most important federal environmental legislation and show you how your representative and senators voted (*lcv.org/scorecard*).

• There are conservation voter leagues in 30 states. Each puts out scorecards and makes endorsements. (Get info at *lcv.org.*) For tips on finding info in the other 20 states, and for local candidates, contact *50simplethings.com/vote*

Step 2. Endorse a candidate. Person-to-person is the best way to communicate political support. So write Letters to the Editor and send e-mail to family and friends. Make the connection between the environment and other issues people care about. Energy use, for example, isn't just about being "green" anymore—dependence on foreign oil makes it a national security concern, and creating green jobs by retrofitting buildings is all about the economy. Get tips and suggestions at *50simplethings.com/vote*

Step 3. Volunteer with LCV or an environmental group. Grassroots groups play a big role in U.S. politics. It's been proven that door-to-door canvassing is the best way to turn out voters, and the most important way to get people to pay attention. Local elections are often decided by a handful of votes; knocking on a few more doors might make all the difference.

Step 4. Work for your candidate. Chances are, they need your help...or your contribution, if you prefer to donate. For a complete how-to guide: *50simplethings.com/vote*

...by President Bush in 2005 to oversee enforcement of the Endangered Species Act.

8. THE LAW OF NATURE

*The Endangered Species Act is the first law in the history
of the world that commits a nation to protecting every
animal and plant species within its borders.*

BACKGROUND. In the early 1970s, America was shocked
to learn that the bald eagle, our national symbol, was on the
verge of extinction.

Suddenly, it became clear to the public just how threatened
wildlife had become. Scientists were able to name hundreds of
plants and animals that might disappear. So the Endangered Species
Act (ESA) was created. This bipartisan law, proposed by a Republi-
can president and passed by a Democratic Congress in 1973, is
something we can all be proud of. The problem is, it hasn't always
been enforced...and powerful lobbies are trying to destroy it.

DID YOU KNOW

• The ESA protects *endangered* species (on the verge of extinction
now), and *threatened* species (headed for extinction). About 1,200
species are officially listed as threatened or endangered, but scientists estimate that more than 6,500 species in the U.S. are actually
at risk of extinction.

• The Endangered Species Act has been successful in protecting
hundreds of species from extinction, including the gray wolf, grizzly
bear, Pacific salmon, and gray whale.

• No species is "expendable"—each has a part to play in the
ecosystem. For example, the reintroduction of the gray wolf to Yel-
lowstone National Park helped control the elk population...which
increased willow and aspen growth along streams...which provided
habitat for birds and cooled water for trout. The thriving trees
meant that beavers could build more dams...which created marsh-
land habitat for ducks, otters, mink, and moose.

• The ESA doesn't just protect species—it also protects their critical habitat, a feature that makes it our strongest and most important environmental law. Preserving habitat, however, cuts into
profits of mining and lumber companies, agribusiness, and land
developers. Irresponsible businesses, and the politicians they give
money to, are constantly trying to weaken the Act.

Success: The bald eagle was finally removed from the list of threatened species in June 2007.

WHAT YOU CAN DO

Your Partner: The Endangered Species Coalition is a national network of hundreds of organizations working to protect endangered species and habitat. Get to know them at *stopextinction.org*

Your Goal: Protect and strengthen the Endangered Species Act, the safety net for our nation's wildlife, fish, and plants on the brink of extinction.

START SIMPLE

Adopt a species. Pick an endangered species you care about and study its life, its habitat, and the threats to its survival. There are thousands of amazing animals and plants on the brink of extinction. Learning about a specific species makes it easier to figure out what you can do to help: *stopextinction.org/endangeredspecies*

STEPS FOR SUCCESS

Step 1. Learn the law of nature. It's hard to defend the law if you don't know what's in it, but reading the entire ESA is a bit much to ask. Solution: Read this excellent Citizen's Guide prepared by the Coalition and Earthjustice: *stopextinction.org/citizensguide*. And support strong endangered species protections by signing the *Endangered Species Act Legacy Pledge* at *stopextinction.org/legacypledge*

Step 2. Big day. Congress has designated the third Friday in May as *Endangered Species Day*, to celebrate our commitment to protecting wildlife and wild places. There are events at libraries, schools, parks, etc., and you can volunteer to help. Find out what's happening in your area: *EndangeredSpeciesDay.org*

Step 3. Inspire a child. Get the Coalition's curriculum for children and make any day Endangered Species Day with your own kids. Or use the curriculum to get local schools involved. Be creative and have fun! Info at *EndangeredSpeciesDay.org*

Step 4. Take a stand. Hundreds of groups are working to protect endangered species across the country. Join one from the Coalition's member list to protect wildlife and habitat in your area.

Bonus: The Coalition has a toolkit that will show you how to help protect species under the ESA. Find it at *stopextinction.org/activist_toolbox*

For more resources: *50simplethings.com/ESA*

Sad truth: More than 90% of the world's tiger populations disappeared in the 20th century.

9. THE RIGHT TREE IN THE RIGHT PLACE

Planting 50 million more shade trees in California cities can provide energy savings equivalent to seven 100-megawatt power plants...if they're on the east and south sides of buildings.

BACKGROUND. You've heard it many times: One of the best things you can do for the environment is to plant a tree...or, even better, a whole bunch of trees. That's because trees are amazing pollution-fighters, water-savers, and soil-savers. They're home to living creatures. They help save energy.

But "planting trees" doesn't just mean randomly putting seeds or saplings into the ground. To get the most from a tree, you need to know what you want from it...and which species will do the job best. Then you need to learn how to take care of it. It's a commitment. But if you're willing to put in the time, planting and caring for trees can be one of the most rewarding experiences you'll ever have.

DID YOU KNOW

• Each year, *a single mature tree* can absorb 48 lbs. of greenhouse gases, help replenish groundwater by returning 1,000 gallons of stormwater to the soil, and supply enough oxygen for two people.

• But just planting a tree doesn't guarantee these benefits—it has to survive long enough. For example: The average downtown urban tree lives about 8–15 years. But trees don't absorb much carbon until they're 10 (some hit their peak at 30; others live to be more than 80). That's why caring for a tree is as important as planting it.

• For maximum benefits, it also matters *where* a tree is planted. According to the U.S. Forest Service, a tree can help save over 50% of a home's daytime air conditioning costs...but only if it's planted on the south side of the building.

• The type of tree is important, too. Pine trees absorb a lot of water, which can be destructive in dry areas. A mature oak tree in a park or yard is ideal urban habitat because its canopy can spread up

In one year, an acre of mature trees absorbs the amount of CO_2 you produce by driving 26,000 miles.

to 100 feet. (However, an oak's rambling roots and branches could be destructive to a city street or small lawn.) A sugar maple planted on a roadway can remove heavy metals like cadmium and lead from the environment.

WHAT YOU CAN DO

Your Partner: TreePeople, an L.A.-based group that has been growing the urban forest and building stronger communities since 1973. Learn more about them at *treepeople.org*

Your Goal: Improve your community's environmental health by planting the right tree in the right place…and taking care of the trees you plant.

START SIMPLE

• **Adopt a tree.** Choose a tree in your yard, or on your street, and spend some time with it. Study it for a week. Consider that you're breathing out CO_2, which the tree absorbs, and that the tree emits oxygen, which you breathe in. You need each other!

• **Look closer.** Learn what's going on "under the surface" of your tree with the *Tree Guide* at *50simplethings.com/tree*

STEPS FOR SUCCESS

Step 1. Do an audit of your community. Get a sense of how trees are needed where you live, learn, work, play, and pray. Research the kinds of trees that would work best in each place. For tips, start at *50simplethings.com/tree*

Step 2. Plant some trees. Do you want to plant at home or help someone with their tree planting project? It's easier to partner with a local tree-planting group, a garden club, Scouts, or 4-H'ers. If you decide to plant on your own, you'll want to do it the right way. Learn how to get started at *treepeople.org*

Step 3. Create a tree project in the community. You don't have to be a tree expert—just someone who wants to lead a project. You'll get plenty of guidance, including information about making a commitment to care for the trees you plant. TreePeople offers a free downloadable copy of *The Simple Act of Planting a Tree: treepeople.org/simpleact*

California study: For every 1,000 trees, stormwater runoff is reduced by nearly 1 million gallons.

10. THE COOL CITIES CAMPAIGN

Chicago's city hall has a 22,000-square-foot garden on its roof, an experimental effort to reduce the "heat-island" effect of urban environments.

BACKGROUND. Let's face it: If you spent too much time worrying about how huge the global warming problem is, you could easily get so overwhelmed that the only action you'd want to take would be jumping back in bed...and pulling the covers over your head.

That's why it's important to come up with new and creative ways of dealing with it—like Sierra Club's Cool Cities campaign, "solving global warming one city at a time." The idea is simple: Citizens work with their mayors to reduce greenhouse gas emissions, using strategies like energy efficiency, renewable energy, and green fleets. All it takes is people like you, who want to make a difference.

DID YOU KNOW

In 2005, Seattle Mayor Greg Nickels decided that even if the U.S. wouldn't sign the Kyoto Protocol (an international global warming treaty), he'd commit to cutting his city's carbon footprint...and invite other mayors to join him. He created the *U.S. Mayors Climate Protection Agreement*, the basis for the Cool Cities campaign.

• Since then, more than 800 Cool Cities have signed the agreement, with more mayors making the same commitment every day.

—Warwick, Rhode Island, cut 1,200 tons of CO_2 emissions just by trading incandescent light bulbs for energy-efficient *light-emitting diodes* (LEDs) in their traffic lights and crosswalk signals.

—Evanston, Illinois, made a commitment to purchase 20% of their electricity from renewable energy sources, including wind.

—Omaha, Nebraska, schools got a grant to retrofit school buses with pollution scrubbers and convert them to biodiesel.

—Seattle cut 2,400 tons of CO_2 by improving their vehicle fleet, including replacing older cars with hybrid-electric cars.

By using LED traffic lights, New York City saves $6 million a year in energy and maintenance costs.

WHAT YOU CAN DO

Your Partner: Sierra Club, made up of 1.3 million members and friends. America's oldest and largest grassroots environmental organization.

Your Goal: Collaborate with members of your community to cut your city's greenhouse gas emissions. Sierra Club's Cool Cities program is designed to help you.

START SIMPLE

Explore the Cool Cities Web site. Register and search for your city on the map. Get involved with an existing campaign, or download the Activist Toolkit to start your Cool Cities campaign: *coolcities.us*

STEPS TO SUCCESS

Step 1. Get a core group together. The Cool Cities campaign is a community effort, so start by getting together with friends and neighbors. It could be a local Sierra Club group, a PTA, a garden club, a congregation, or any business.

Step 2. Get the mayor to sign on to the U.S. Mayor's Climate Protection Agreement. First, research your mayor's positions on the issues; then see what positive things are already happening. You can tailor your pitch and encourage your mayor to build on current successes. Set up a meeting and be sure your group represents a good slice of the community. The Activist Toolkit includes step-by-step guidance, including advice on approaching a reluctant mayor.

Step 3. Prepare a citywide "emissions inventory." It offers insight into where CO_2 is coming from and how you can reduce it. Sierra Club's Cool Cities program can direct local governments to groups that help conduct these inventories.

Step 4. Develop a local climate action plan. Create a Green Ribbon Commission dedicated to developing a climate action plan—the long-term strategy for reducing your city's carbon footprint. It can include representatives from the city as well as local citizens. To help you with the process, many Cool Cities have their climate action plans available online. Links to many of these plans are at *coolcities.us*

For more resources: *50simplethings.com/coolcities*

In St. Paul, MN, a power plant uses the heat created by producing electricity to heat local buildings.

11. THEY PAVED PARADISE...

The U.S. has 3.9 million miles of paved roads, enough to circle the Earth at the equator 157 times.

BACKGROUND. Next time there's a rainstorm, go outside and really watch. Look at all the water flowing off your walkway, your roof, and your driveway, and rolling down the street. Watch it literally go down the drain.

That water is called *runoff*, and as it moves toward the sewers it picks up oil, pesticides, garbage, and dirt particles called sediment. You may think these pollutants and garbage are magically removed from the water before it gets to rivers and streams, but they're not. The gross stuff you see (and *don't* see) going down the drain, the "oily rainbow," is washing right into our waterways.

DID YOU KNOW

• In a natural system, rainwater lands on the Earth and slowly seeps into the ground. This replenishes aquifers and filters impurities out of the water.

• Today, too much rainwater never *touches* the dirt because we've covered it with impervious surfaces—roads, driveways, roofs, parking lots. The area in the U.S. devoted to paved roads and parking lots alone covers an estimated 61,000 square miles—an area larger than the state of Georgia.

• The result: stormwater runoff. In fact, polluted runoff is the #1 source of water pollution in the U.S. Beach closings, fish kills, toxic algae blooms, and rivers choked with mud all result from uncontrolled stormwater runoff.

• In 1972, Congress passed the Clean Water Act, promising to eliminate water pollution and ensure that our waters are "fishable and swimmable." More than 35 years later, nearly half of America's lakes, rivers, streams, and coastal waters remain polluted. Industrial water pollution has been reduced, but polluted runoff is still largely uncontrolled. It continues to poison our waterways.

Every year, 2.2 million acres of agricultural property and forests are converted to suburban and...

WHAT YOU CAN DO

Your Partner: Waterkeeper Alliance. On more than 170 waterways around the world, local Waterkeepers are on patrol, standing up to polluters and enforcing our right to clean water. Waterkeeper Alliance supports these grassroots advocates. See: *waterkeeper.org*

Your Goal: Reduce the amount of runoff polluting our waterways, starting with the runoff that comes from your own home.

START SIMPLE

• *Keep it clean.* Waterkeeper's "4 things anyone can do to help keep stormwater clean": 1. Wash your car on the lawn or at a commercial car wash, not in the driveway or street. 2. Compost or recycle yard waste. 3. Sweep—don't hose—driveways and sidewalks. 4. Use fertilizers and pesticides sparingly, if at all, and properly dispose of them and other household chemicals.

STEPS FOR SUCCESS

Step 1. Set up a rain barrel at your house. A rain barrel collects water that runs down your gutters and can then be used for watering your lawn and other household tasks. If you're a renter, ask your landlord about putting up a rain barrel. Find tips at *waterkeeper.org*

Step 2. Create a rain garden. It directs runoff across your yard and into an area with lots of native plants. "Holding the first inch of water from a storm keeps 90% of pollutants and nutrients out of the local waterway," says Jeff Odefey of Waterkeeper Alliance. "Gardens are better filters than lawns. A garden can absorb seven inches of water per hour, grass only two." Get info at *waterkeeper.org*

Step 3. Join a local Waterkeeper organization or other watershed group. This will help you and others to become better advocates for stormwater cleanup in your community.

Step 4. Do a demonstration project for your community. With your group, install a rain garden or other project as a demo to help educate your community. For suggestions: *50simplethings.com/runoff*

Step 5. Become an advocate. Construction runoff is a major part of the problem. Learn as much as you can about this issue; try to convince local builders of its importance, and explain the alternatives. Check out *50simplethings.com/runoff* or *waterkeeper.org*

...urban land uses, resulting in at least 80 million tons of sediment entering our nation's waterways.

12. SAVE YOUR ENERGY!

Heating and cooling systems in the U.S. together emit more than 150 million tons of CO_2 into the air each year.

BACKGROUND. Of all the different strategies we can use to tackle global warming and air pollution, energy efficiency is the quickest, cheapest, cleanest, and easiest. It costs nothing, for example, to turn off lights and electronics when you leave a room. Sealing leaks with weather stripping and caulking is cheap—especially compared to what it saves in fuel and money. Even the energy-savers that cost a little more up front, like compact fluorescent lightbulbs and appliances with the ENERGY STAR label, provide a financial payback—while they significantly cut CO_2.

So why isn't energy efficiency a national priority? How can we expect to win the hard fight to stop global warming if we're not even committed to doing the *easy* things? We have to work together to change this, starting where we live...but we also have to make sure that energy efficiency becomes a priority for business and government. America needs to put its energy into *saving* energy.

DID YOU KNOW

• Energy is the largest industry in the world. More than 80% of the world's air pollution and greenhouse gas emissions are caused by energy production and use.

• Much of the energy we use literally goes out the window. The gaps you'll find around the windows and doors of an average American house add up to the equivalent of a wide-open 3' x 3' window.

• Market testing shows that since President Carter appeared on TV in 1977 urging us to turn down the heat and wear sweaters, Americans have equated saving energy with sacrifice and deprivation.

• But being energy efficient and using energy-efficient technologies is really just being smart. For example: Students in an upstate New York school district, working with the Alliance to Save Energy's Green Schools Program, noticed the lights were on in their school lunchroom during the day—even though the space had windows on three sides. They simply turned off the lights and saved $11,000 in energy costs that year for their school district.

Microwaves use about 66% less energy than conventional ovens.

WHAT YOU CAN DO

Your Partner: The Alliance to Save Energy (the Alliance) promotes energy efficiency worldwide to benefit consumers, the environment, and the economy. Get to know them at *ase.org*, and get environmental and money-saving tips at *ase.org/consumers*

Your Goal: To look at your daily energy decisions through the lens of their environmental impact...and inspire others to do the same. Let's make energy efficiency a national priority.

START SIMPLE

Take the challenge: Multiply your environmental impact. The Alliance has a program to spread the efficiency message called the "6° of Energy Efficiency Challenge." The simple idea: Test your "Energy IQ" with six quick questions, take 1–6 energy-efficiency actions, encourage six friends to do the same...and learn about our 6° of interconnectedness at *sixdegreechallenge.org*

STEPS FOR SUCCESS

Step 1. Conduct or get a home energy audit, and then follow its recommendations. The average U.S. home creates nearly twice as much CO_2 emissions as the average car, so it's a good place to start. Find hidden leaks and take care of unseen problems that make a difference in comfort, cost, and energy. Your local utility may offer free (or low-cost) audits. We've got links to make the whole process easier at *50simplethings.com/energy*

Step 2. Teach energy efficiency. The Alliance's multidisciplinary, hands-on Green Schools Program integrates lessons about energy with other subjects—like math. Green Schools works with school districts, helping teachers, students, and facility managers work together to save schools energy and money. "If you can build energy efficiency right into the educational system K–college, students will understand energy issues, and influence their parents—and maybe they can even transform the way we make choices about energy," says Alliance Education VP Merrilee Harrigan. Check their Web site for K–12 lesson plans to download: *ase.org/educators*

Step 3. Make it a policy. Help spread the efficiency message; join the 50,000 grassroots advocates throughout the U.S. who are part of the Alliance's Action Network. Sign up at *ase.org/policyaction*

Heat loss through your basement floor can account for as much as 20% of your total heating bill.

13. THINK GLOBALLY, EAT LOCALLY

There are 5 million fewer farmers in the U.S.
today than there were in the 1930s.

B ACKGROUND. Ever take a look at where the food in your
kitchen comes from? It's pretty amazing. Depending on the
season, you might have oranges from Israel, grapes from
Chile, apples from New Zealand.

It's an impressive distribution system…but people are starting to
notice that there are hidden environmental costs in routinely trans-
porting food across a continent or halfway around the world. The
result is a new way of thinking about what we eat, known as the
"local foods" movement.

DID YOU KNOW

• In the U.S. today, most "fresh" food is trucked long distances
from huge farms, many of which are corporate-owned. A head of
romaine lettuce typically travels 2,055 miles from farm to store; a
stalk of celery averages 1,788 miles; an onion, 1,675 miles.

• In fact, according to the Natural Resources Defense Council
(NRDC), the typical American meal contains ingredients from at
least five countries outside the United States.

• Imported food is transported by planes, ships, and trucks. "In
2005 alone," NRDC reports, "total agricultural imports into California resulted in almost 250,000 tons of global warming pollution,
more than 6,000 tons of smog-forming nitrogen oxides and 300
tons of sooty particulate matter."

• Another important environmental consequence is the impact on
farmland and our food system. Buying food from far away puts local
farmland (and farmers) at risk. It takes away the incentive to protect this valuable land, and that's one reason we're losing so much
of it. According to a recent study, the U.S. is losing two acres of
mostly prime farmland *every minute* to development.

Movie Night: Rent *The Real Dirt on Farmer John,* a great documentary about a farmer named John.

WHAT YOU CAN DO

Your Partner: The Food Routes Conservancy (FRC), a national nonprofit "dedicated to reintroducing Americans to their food." Check them out at *foodroutes.org*

Your Goal: Support farmers in your area by eating locally grown food. FRC will help you find local food, join a local *Buy Fresh Buy Local* chapter, and spend some time on a farm discovering just how your food is produced.

START SIMPLE

Go on a treasure hunt. See if you can find your local farmers market(s), food co-op, or a U-pick farm stand. *Bonus points:* Look for "producers-only" farmers markets to assure local sources.

STEPS FOR SUCCESS

Step 1. Take the "Buy Local" challenge. Spend $10 a week for the next year on local foods. Sign up at *foodroutes.org/buy-local-challenge.jsp.* They'll show you how to find seasonal foods grown by local food producers, as well as stores that carry them in your area.

Step 2. Join a CSA (Community Supported Agriculture), which lets you buy a box of local, in-season produce once a week and have it delivered. If one box is too much for you and your family, split it with others. In 1990, there were only about 50 CSAs; now there are more than 1,000. For a Web site listing every CSA in the U.S.: *foodroutes.org*

Step 3. Visit local farmers (great for kids!). There's a lot to learn about where your food comes from and how different (and how much better) it is when it comes from local sources. The best educators about Local Foods are the people who grow it. To find out more about farm tours, check *50simplethings.com/localfoods*

Step 4. Host a local foods party. Print out the recipes and where you got each ingredient. Invite a local farmer to be your guest.

Step 5. Organize an Eat Local Challenge Week. This great idea is spreading across the country. For a week, stage events that use and promote local foods in your community. Restaurants may offer local food entrees; grocers can have specials on local produce. Get detailed instructions from us at *50simplethings.com/localfoods*

By increasing your local spending just 1%, you can make a 5–10% difference in local farmers' incomes.

14. GOD'S GREEN EARTH

More than half (55%) of the Americans who consider themselves religious now favor stronger environmental regulation.

BACKGROUND. Is protecting the environment a religious issue for you? Do you see it as a way of honoring God's creation and serving humankind?

If you do, you're part of an important movement: faith-based environmentalism. "Loving God and loving God's creation means protecting the world around us," explains the Reverend Sally Bingham of San Francisco's Grace Cathedral.

As global warming gets more intense and threatens more of Earth's creatures, many congregations are being called to respond. If you'd like your *own* congregation to join the fight, you've got a powerful ally. No, not *that* ally. We mean Interfaith Power and Light, your partner in this section.

DID YOU KNOW

• Activism is part of the American religious tradition. Clergy were an important force in the abolitionist and civil rights movements.

• Religious Americans have the numbers to make a big difference. According to Gallup, over 65% of Americans say they attend church or synagogue. That's more than 200 million people.

• Interfaith Power and Light (IPL), an organization dedicated to mobilizing people of faith against global warming, showed *An Inconvenient Truth* to more than half a million people in congregations across the U.S. in 2006. Regional IPL chapters develop programs independently. The Illinois IPL, for example, has helped 12 congregations purchase energy from a local wind farm.

• Leaders of many faiths are lobbying the government to take action. In 2007, Christian, Jewish, and Muslim leaders signed a letter to President Bush and Congress calling global warming a "moral and spiritual issue" and urging them to curtail greenhouse gas emissions and invest in renewable energy.

A new Web site offers congregation members energy-efficient products at discount prices: *shopipl.org*

WHAT YOU CAN DO

Your Partner: Interfaith Power and Light. Founded in 2000, IPL promotes energy conservation within religious organizations, and can help you develop programs for your congregation. Get to know them at *interfaithpowerandlight.org*

Your Goal: Become an advocate for change in your congregation by sharing the message that climate change is a moral issue.

START SIMPLE

• **Movie night.** Screen *Lighten Up! A Religious Response to Global Warming* at your congregation. This 20-minute Emmy-nominated film is about faith and environmental stewardship. Organize a group discussion after the showing. Find out how to get a copy of the film and obtain a facilitator's guide to discussion at IPL's Web site.

• **Get the guide.** The EPA's *ENERGY STAR for Congregations* program includes a downloadable step-by-step guide to analyzing and upgrading your congregation's facilities. Get it at *50simplethings.com/IPL*

STEPS FOR SUCCESS

Step 1. Find out how green your church is. Talk to your religious leader about the connection between religion and our role as stewards of creation. Ask him or her to consider delivering a sermon on the topic. Then talk to the members of your congregation and try to get a sense of who might be willing to be part of a congregation-wide effort.

Step 2. Meet IPL. Bring information about IPL to your congregation's leadership. Introduce IPL's six-point covenant, which outlines actions your congregation can take. 4,000 U.S. congregations are already a part of the campaign. Find it at *interfaithpowerandlight.org*

Step 3. Join 'em. IPL had 26 state affiliates at last count. If your state is among them (check it out at *interfaithpowerandlight.org/state*), encourage your congregation to join. If it's not, create an IPL chapter. Guidelines are at *interfaithpowerandlight.org/guidelines.htm*

Bonus: Calculate your congregation's carbon footprint. Visit IPL's Web site and download materials on how to become a Cool Congregation.

Check out these related Web sites: *www.creationcare.org*, and *www.nrpe.org*

15. UP THE CREEK...

Sacramento County, California, has held its annual Creek Week for 18 years. Hundreds of volunteers join in to clean the area's creeks, play games, make art from the collected trash, and celebrate with a barbecue.

BACKGROUND. On a hot summer day, it's nice to take off your shoes and wade in a local stream...unless there are signs posted telling you the water is unsafe. That's a jarring image, but it's not all that uncommon. According to the EPA, about 40% of all streams and rivers that have been tested in the U.S. aren't safe for activities like fishing and swimming.

Actually, there are a lot of things threatening our streams—from garbage to fertilizer runoff (see p. 40), sediment that blankets stream bottoms to cement barriers that "straighten" them out. The good news is that many communities across the nation have their own stream-protection groups. Why not join them? It's a great way to preserve a waterway, educate kids about the environment, bring the community together...and have fun.

DID YOU KNOW

• For more than 15 years, community volunteers and students from 23 schools in central New York have been wading into local streams several times a year to determine the health of the water. They look at the bugs living in streams and test the water's chemistry through Project Watershed, an Izaak Walton League of America (IWLA) Save Our Streams project.

• To help clean up local streams, members of the Tennessee IWLA chapter invented the "litter skimmer," a device made of empty milk jugs and other recycled materials. They stretch it across a slow-moving area in a stream to capture floating trash, then local volunteers empty it periodically. Thousands of bags of trash have been removed from Tennessee creeks and streams this way. Find out how to make and use a litter skimmer at *iwla.org/sos/litterskimmer*

• In Hamilton, Ohio, at their annual Halloween festival, the local IWLA chapter collects buckeyes and other native nuts donated by the event's more than 1,000 participants. Volunteers propagate the appropriate ones for seedlings, which are then used for streamside plantings to help reduce stream bank erosion.

In France, scientists are using trout to test stream health...

WHAT YOU CAN DO

Your Partner: The Izaak Walton League of America (IWLA). Since 1922, IWLA has protected America's outdoors through community-based conservation, education, and outdoor recreation.

Your Goal: Clean, restore, and protect streams in your community.

START SIMPLE

Take a hike. Go for a walk along a local creek with your family. Make it more fun: First, download the Izaak Walton League's one-page Stream Guide and take it along. Five things to look for: Water clarity, insect life, healthy vegetation, fish, and wildlife (including birds, lizards, snakes). To get the guide: *iwla.org/sos/streamguide*

STEPS FOR SUCCESS

Step 1. Throw a party! On a nice summer day, gather friends and family for a *Trash-Pickup Barbecue Party*. Bring rubber boots, gloves, trash bags…and the barbecue. Choose a section of the stream to clean up. Offer a prize for "most unusual trash."

Step 2. Form a "Friends of the Stream" group. Get friends, family, church, 4-H, or a Scout group to join you. Become the caretakers for a stretch of the creek; IWLA's Save Our Streams (SOS) program can teach you how to monitor, clean, and restore the stream. Adopting a stream will make your community more attractive, improve habitat for wildlife, and restore a natural resource. Find everything you need to start a stream monitoring or restoration project at *iwla.org/sos*. For info on ordering SOS videos and publications, go to *iwla.org/cleanwater/publications*

Step 3. Inspire children. Meet with teachers and administrators to set up a "healthy creek" project. Students can perform tests to determine water quality, study the effects of pollution on watersheds, and perform hands-on habitat restoration. Find links to teacher guides at *50simplethings.com/streams*

Step 4. Get your town involved. Report your progress to the mayor or city council. Ask them to sponsor a communitywide Creek Week. Local businesses can offer prizes, drinks, and snacks, or clean-up equipment like dumpsters. Invite water and wildlife experts to speak, and have hands-on educational activities for kids. Then dive in and start cleaning.

…because the fish can detect *one-billionth* of one gram of pesticides in a liter of water.

16. GROWING WARMER

According to a study by Stanford University scientists, one of every five wildflower species could die out over the next century if levels of carbon dioxide in the atmosphere double, as expected.

BACKGROUND. If you're a gardener, you might know something a lot of people don't: Global warming is already changing nature. Maybe you've noticed that plants are leafing out and blooming earlier…or that birds and butterflies are breeding and migrating earlier…or that "new" birds are showing up at your backyard feeder

You might have thought, "Oh, it's just a warmer winter." But there's much more than that going on, and scientists predict additional, permanent changes. As a gardener, you're a steward of the environment. So it's important to be aware that there are simple, thoughtful things you can do that will make an enormous difference in your own backyard, and in your community.

DID YOU KNOW

• Climate change is already visible. Maps of "hardiness zones," which show the U.S. areas where specific plant types grow best (found on the backs of most seed packets), have had to be updated twice in the past 20 years to account for rising temperatures.

• The maps tell us that because it's getting hotter everywhere, plant habitats are moving north. "You could say D.C. is the new North Carolina," says the curator of the U.S. Botanic Gardens.

• Scientists also suggest climate change may be so intense that before the end of this century, many states will no longer have favorable climate for their official state tree or state flower.

• Gardeners have the power to make a difference: People in more than 43 million U.S. households consider themselves gardeners. And an estimated 91 million households participate in "lawn and garden activities," spending more than $35 billion annually.

• According to *English Nature* magazine, "Gardeners can help mitigate the effects of climate change by providing habitats for the most threatened species, and/or saving water by changing gardens to use plants which require less."

Looking for native plants? You'll find a directory at *plantnative.org*

WHAT YOU CAN DO

Your Partner: National Wildlife Federation (NWF), dedicated to combating global warming, connecting people with nature, and protecting wildlife for our children's future since 1936. Visit *nwf.org*

Your Goal: Adopt sustainable gardening techniques that help minimize global warming.

START SIMPLE

• **See the light!** Switch your outdoor lights from incandescents to CFLs, and landscape lights to solar or other energy-savers. OK, it's not gardening, but it's an easy way to start changing things outside.

• **Knowledge is power.** Go to *www.nwf.org/gardenforwildlife/* and download NWF's *Gardener's Guide to Global Warming.* Sign up for NWF's newsletter *Wildlife Online;* find out more about global warming, gardening, and how you can make a difference for wildlife.

STEPS FOR SUCCESS

Step 1. Garden to save water. If you're water-conscious when you plant, your garden will flourish in dry seasons. Grow drought-tolerant plants. Use drip irrigation or soaker hoses. Spread mulch around plants to keep water close to roots. Use a rain barrel (see p. 40). No matter where you live, saving water makes gardens and the Earth healthier. Check out *www.nwf.org/gardenforwildlife/*

Step 2. Go native! Many experts recommend native plants that are adapted to the rainfall in your area. As conditions change, they'll adapt better. Planting natives also improves wildlife habitat and brings birds into your yard. For info: *www.nwf.org/backyard/food.cfm*

Step 3. Cut the grass. Think about reducing the size of your lawn. Every year, Americans put nearly 80 million pounds of pesticides on their lawns. Those pesticides affect native plants and wildlife, and leach into the groundwater. Lawns also account for 32% of outdoor water use—smaller lawns save water and gasoline (power mowers). Find out about "green" lawns: *www.nwf.org/gardenforwildlife/*

Step 4. Grow a movement. Look for ways to change your community's values. For example, many homeowners' associations insist on a certain size lawn—and impose fines on residents who don't comply. Speak out about this kind of unsustainable landscaping. We can change people's perceptions about lawns and gardens—learn how at *50simplethings.com/garden*

In one acre of land, there can be more than one million earthworms.

17. GROW A GREEN MARKETPLACE

"Walk the aisles at your local supermarket or big-box retailer. How many of the products you see reflect sustainability values? How about the companies that make them? How about the stores that sell them? How many shoppers are bothering to ask such questions?"
—*Joel Makower, in his blog "Two Steps Forward"*

BACKGROUND. When you hear the term "green products," you probably think of things like energy-efficient lightbulbs and environmentally friendly cleaners. They're important, of course...but new, improved consumer items aren't the only way to "go green." Building a green marketplace also means learning to look at the things we already have in a new way.

When you see an attractive $10 cotton shirt on a rack in a department store, for example, does it seem like a good deal? Sure, it doesn't cost *you* much, but what did it take to make that shirt? How much water and energy—and pesticides—did it take to grow the cotton? What extraction methods were used to mine the metals for buttons or zippers? Was it produced in a developing country by someone making $1 a day in a factory emitting tons of pollution... and then shipped here, using oil from the Middle East? If you look at it *that* way, it's an expensive shirt.

This is not to say that everything affordable is ecologically unfriendly, or that even if it is, we shouldn't buy it. It's just that we need to change the way we think about what we buy. We need to begin thinking about the *real* cost—not just what it costs us, but also what it costs the Earth.

DID YOU KNOW

Here's what can happen when people think about things in a new way:

• JK Rowling, author of the *Harry Potter* series, got her publishers to print her books on postconsumer recycled paper, spawning a trend that has seen an additional 300 publishers, including Harper-Collins UK and Random House, adopt green paper policies.

• Clorox launched its first new brand in 20 years: Green Works, a

In 2007, General Electric intoduced the Earth Rewards credit card. What's the appeal?...

line of biodegradable cleaning products made from coconuts and lemon oil, packaged in recyclable bottles, and not tested on animals. It's endorsed by the Sierra Club.
• Spurred by green consumers, sales of Fair Trade Certified™ products grew to $2.3 billion in 2006. Costco, Sam's Club, Dunkin' Donuts, and McDonald's have all begun carrying Fair Trade coffees.

WHAT YOU CAN DO
Your Partner: Co-op America. Since 1982, their mission has been to harness our economic power to create a socially just and environmentally sustainable society. Get acquainted at *coopamerica.org*

Your Goal: Use your buying power to transform our consumer economy so it's ecologically sustainable.

START SIMPLE
Read the label. Look out for *Greenwash*—false environmental claims that businesses make about their products. Before you believe them, research products and producers. Start with Co-op America's *responsibleshopper.org*, then check out *50simplethings.com/market* for links to articles and reports.

STEPS FOR SUCCESS
Step 1. What's "green"? We can't create a green marketplace until we know what *green* means. Is it enough to be made from renewable materials like bamboo or corn...or do those materials have to be grown sustainably and locally? How important are the ethics of the manufacturer? Is price a factor? Each of us has to decide. Tell us what *you* think at *50simplethings.com/market*

Step 2. Category thinking. Co-op America's Alisa Gravitz notes, "For some people, 'going green' can sound too challenging; they wonder how they're going to make big changes in their lives...and then they don't take any steps at all." Her solution: Adopt *category thinking*. Pick one category of consumption for a year—like energy use, food, or clothing—and take green steps within that category, a little at a time. We can help: go to *50simplethings.com/market*

Step 3. Vote with your dollars. Once you know what you want, talk to store managers about putting more green products on the shelves. If you can't find it locally, buy from online sources like Co-op America's National Greenpages™ at *greenpages.org*

...It will invest 1% of consumer purchases made with the card in carbon offset projects.

18. PARK IT HERE

The total area of urban parkland in the United States is more than 1 million acres, and more people use urban parks than national parks. In fact, New York's Central Park gets about 25 million visits annually —more than five times as many visits as the Grand Canyon.

BACKGROUND. Urban green spaces—natural areas preserved in urban settings—are an essential part of city life. Some are tiny oases surrounded by concrete. Others are huge ecosystems, like Central Park. Whatever their size, urban parks are beloved for the benefits they provide, like recreation and refuge.

But most city dwellers don't realize that under the surface—literally—green spaces are also working 24/7 to support life. In a way, they're the unsung protectors of the city environment. They provide habitat for local animals and rest stops, or "flyways," for migrating birds. They capture and filter rainwater, reducing polluted runoff (see p. 40). They help combat the urban heat-island effect, cooling city air through transpiration and evaporation. City green spaces are invaluable assets...which is why it's important to preserve and protect them, and to create more in our communities.

DID YOU KNOW
• Urban green spaces can play a *big* role in fighting global warming. According to an authoritative 2007 study, it would take only 10% more green space in cities to "reduce urban surface temperatures by as much as 4°C—which is equivalent to the average predicted rise that global warming will cause by the 2080s."

• Trees are also effective in fighting urban runoff. American Forests, a nonprofit that works to protect trees, estimates that park trees have saved us $400 billion that would have been spent building stormwater retention facilities. Yet tree cover has declined by as much as 30% in many cities over the last several decades (see p. 36).

• People across the nation (and around the world) have formed independent "Friends of Parks" groups dedicated to maintaining, improving, and creating green spaces in their towns. In 1975, for example, a small group of concerned citizens founded Friends of the

Good investment: Washington D.C.'s nearly 30% tree canopy coverage removes...

Parks to improve Chicago's urban parks. To date, they have saved more than 100 acres of lakefront parkland from private development and prevented the loss of 30 more acres of neighborhood parkland to private and municipal interests.

• Cities are reclaiming "brownfields"—abandoned industrial or municipal areas—a turning them into parks. NYC is transforming its largest landfill into Fresh Kills Park, with 2,200 acres of open space. Over half of the site in the 30-year project is already composed of wetlands, open waterways, and unfilled lowland areas.

WHAT YOU CAN DO

Your Partner: The Trust for Public Land (TPL), a national conservation nonprofit that conserves lands and helps create parks in cities. Visit them at *tpl.org*

Your Goal: Support conservation efforts for parks in your community, and help support the creation of more urban green spaces.

START SIMPLE

Be a park detective. Pick a park and download our Park Guide at *50simplethings.com/parks*. Explore your park. Its environmental benefits are hidden—can you find them? They're in the tree filtering out CO_2 and giving oxygen…in the soil that's filtering water…in the habitat where birds and other animals are making their homes.

STEPS FOR SUCCESS

Step 1. Vote for parks. Parks or green spaces often end up as election issues. Join the Conservation Campaign, which works to mobilize public support for the ballot measures: *conservationcampaign.org*

Step 2. Form or join a Friends of the Park group. Learn more about it at *50simplethings.com/parks*

Step 3. Create urban green spaces. Are there enough parks in your city? Explore the possibility of transforming "brownfields" into green space with your parks department. TPL helps communities renew contaminated areas and transform them into parks. See what they do at *tpl.org*. Visit the *Parks for People* section.

Step 4. Get involved in a land trust—a nonprofit that works with landowners to conserve land through acquisition or easement (a sort of conservation right of way). Land Trust Alliance has a list of local trusts. Find one near you: *ltanet.org/findlandtrust*

…some 540 tons of air pollution a year—saving the city $2.5 million.

19. SAVE THE WHALES...AGAIN

The blue whale, the largest creature ever to live on Earth, has been reduced to less than 5% of its original population by commercial whaling.

BACKGROUND. How many times have you seen bumper stickers like "Save the Whales" or "Forget the Whales, Save the Humans" or "Save the Whales—Trade Them for Valuable Prizes!"? The idea of protecting whales was so prevalent during the 1970s and '80s that it has become a cliche.

That was so long ago, you might assume we've already saved the whales and moved on to other things. Unfortunately, you'd be wrong. Whales, dolphins, and other marine mammals still face huge threats—toxic chemicals, intense sources of noise such as military sonar, food shortages, and even increased whaling. That's why there's a new fight to save the whales today. You can help.

DID YOU KNOW

• Although the International Whaling Commission imposed a moratorium on whaling in 1986, Japan, Norway, and Iceland keep hunting them. Since the moratorium went into effect, those three countries have killed more than 27,000 whales.

• The Japanese kill whales, they say, for "scientific research." But they sell the whale meat as food. It's even used for whale-meat dog and cat food.

• What makes this especially outrageous is that whales are already threatened. The U.S. Endangered Species Act (see p. 34) lists eight endangered whales in U.S. waters alone.

• Public opinion can make a difference for marine mammals. Until the early 1990s, dolphins (which are technically whales, too) were routinely killed by tuna fishing—100,000 of them a year. Then consumers joined forces and protested, compelling companies to adopt dolphin-safe policies. Today, as a result, fewer than 1,000 dolphins are killed in tuna nets each year.

The U.S. Navy's use of an experimental sonar, one of the loudest sounds ever generated in the ocean...

WHAT YOU CAN DO

Your Partner: Earth Island Institute's International Marine Mammal Project, a leader in the fight to end whaling and the killing of dolphins. Get to know them at *earthisland.org/immp*

Your Goal: Make a connection with the Marine Mammal Project. Protecting whales is a long-term process, and there aren't a lot of hands-on things to do. But situations will arise where your help is needed—and if you're paying attention, you can get involved.

START SIMPLE

Be a Dolphin Scout. The U.S. tuna market is entirely dolphin-safe by law. To keep businesses honest, Earth Island has a monitoring program that surveys tuna canneries and checks the records of tuna processors and importers. They're asking you to keep an eye out: If you find canned tuna that doesn't have a dolphin-safe logo (some comes in from Mexico), mail them the label and tell them where it's being sold. They'll take action to try to get rid of it.

STEPS FOR SUCCESS

Step 1. Join the Earth Island Whale Watchers. You'll get alerts when legislation comes up or citizen action is required. Pass the info on to friends as you get it: *earthisland.org/immp/50whale*

Step 2. Go on a whale watch. You can't bring the whales to you, so go to them. The sight of these imposing creatures in their natural environment is awe-inspiring. Plus, you'll be supporting safe commerce around whales that gives people a stake in protecting them. Choose a whale-watch outfit that promotes conservation and keeps its distance from the whales to avoid stressing them. If you can't get to the ocean, you can still watch a film. Info is available at *50simplethings.com/whales*

Step 3. Be a whale watcher. Start your own "whale watch." Look for items in the news and post them on your Web site or blog. E-mail alerts to your friends. Stay connected to the effort so you'll be there when it counts. When Earth Island has an opportunity to have an impact on commercial whaling or high-tech sonar, they'll need your help. Public opinion is the most powerful tool they have to change things, so be ready to speak up.

For more resources: *50simplethings.com/whales*

s disrupted the mating, feeding, and migration of whales. It has reportedly caused whales' ears to bleed.

20. TOO MUCH GAS!

By improving the mileage of the American car fleet by only 10 mpg by 2020, as Congress finally did, we'll save a million barrels of oil a day.

BACKGROUND. With global warming, fears about foreign oil, and sky-high gasoline prices, you'd figure America would be rushing to make cars that use a lot less gas, right? Well, we're not. The U.S. has one of the lowest fuel economy standards of any industrialized nation in the world. In fact, environmental groups have had to fight tooth and nail just to get Congress to approve the first increase in gas mileage standards in 30 years.

This seems crazy. It should be obvious that while we're developing alternative fuels (see p. 78) and plug-in electric cars (see p. 20) for the future, we need to dramatically cut down the amount of gasoline we consume *right now*.

DID YOU KNOW

• Every day, America uses about 405 million gallons of gasoline. The result: greenhouse gases, air pollution, contaminated groundwater, oil spills, a national security threat (about 60% is foreign oil). And we pay oil companies *$1 trillion a year* for the privilege.

• In 1975, Congress passed the first official gas mileage regulations, called Corporate Average Fuel Economy (CAFE) standards. The cars and light trucks in every manufacturer's fleet are required by CAFE standards to meet a certain average fuel economy level.

• The law worked. In 1975, average gas mileage for American cars was 12.9 mpg. By 1985, the average mpg had increased to 27.5. CAFE standards were supposed to keep going up. But Detroit pressured the government to relax them in 1986…and 20 years later, passenger car mpg hasn't improved *by even 1 mpg*.

• In 2007, Congress finally increased CAFE standards—by 8 mpg. A victory? Sort of: The government immediately used this modest increase as an excuse to keep states from passing more far-reaching standards. The EPA blocked, for the time being, a California law that substantially tightened emissions standards, saying improved CAFE standards were enough. Apparently, if we want cleaner, more fuel-efficient cars, we have to demand them ourselves.

Top 5 international gas hogs: 1) U.S. 2) China 3) Japan 4) Canada 5) Mexico

WHAT YOU CAN DO

Your Partner: The Natural Resources Defense Council (NRDC), one of our largest and most effective environmental groups. NRDC's *Move America Beyond Oil* initiative is committed to building support for clean, efficient technologies and policies that push them into the market. Join the effort; check early and often for ways to make your voice heard at *beyondoil.nrdc.org/50simplethings*

Your Goal: Help cut global warming pollution by using less gas.

START SIMPLE

It's a gas. Check the gas mileage of your own car. Experiment: See what happens when you do little things. Inflating the tires properly could save 200,000 barrels a day if everyone did it. Going 65 mph instead of 75 reduces gas consumption by 15%. Turning the car off if you have to wait longer than 60 seconds saves up to 19% of your fuel. Check our gas-saving tips at *50simplethings.com/savegas*

STEPS FOR SUCCESS

Step 1. Talk to your local car dealer. Say you're glad that Congress passed higher fuel economy standards, but you know carmakers can and *should* beat those goals. You might get a puzzled look at first, but if they hear from enough people about vehicle performance standards, they'll send a message up the food chain. Dealerships are used to talking with customers about cup holders; they'll be surprised to hear about policy. Get info at *beyondoil.nrdc.org/50simplethings*

Step 2. Send a message to carmakers. Changes in the car-rental market signal what car customers want...so demand fuel efficiency from rental companies; rent only high-mileage cars. And comment on the Internet; many carmakers are just starting to blog, so they're paying attention to what's said on their sites. Your comments about better gas mileage and global warming pollution will have an effect.

Step 3. Raise the issue with federal and state officials. Join with others in your community and set up meetings with candidates, representatives in D.C., the governor, and state legislators; tell them what you think. Stress additional issues like a stronger economy and national security. Hearing from hundreds of thousands of voters about this policy's importance will make a difference. For NRDC's regular updates, plus ideas for saving gas and pushing for better policy: *beyondoil.nrdc.org/50simplethings*

While the world's population doubled between 1950 and 1996, the number of cars increased ten-fold.

21. SUPPORT FAIR TRADE

"Farmers can't take care of the environment on an empty stomach." —Juanita Baltodano, Costa Rican coffee farmer

BACKGROUND. The agricultural regions of Africa, Latin America, and Asia are some of the most ecologically sensitive areas in the world. The people who live and farm there are also some of the world's poorest. These small-scale farmers have traditionally been good stewards of the land, cultivating crops in ways that protect soil and water, and provide habitat for wildlife.

The problem is that the majority of them struggle to put food on their tables, keep a roof over their family's heads, and send their children to school. The only way they can afford to keep protecting the Earth for *us* is if they have a secure market for what they grow or make, and can earn a decent wage at it. By buying Fair Trade Certified™ products, you give them that opportunity.

DID YOU KNOW

• Here's how Fair Trade works: Wholesalers in the U.S. and Europe pay farmers a premium price for their goods. (More than $100 million in *additional* income has gone directly back to these farms since 1998.) The farmers, in turn, agree to uphold certain environmental, economic and social standards. Consumers show their support by looking for the Fair Trade Certified label when they shop.

• In 2006, worldwide sales of Fair Trade Certified products were an estimated $2.3 billion, directly benefiting more than 1.4 million farmers and farm workers in 60 developing countries.

• Environmental standards are a big part of the system. Most Fair Trade products are organic, while the standards prohibit the use of genetically modified organisms (GMOs) and call for Integrated Pest Management (IPM) to reduce chemical pesticides. Sustainable farming methods ensure soil fertility while protecting farmers' health and preserving valuable ecosystems for the future.

• Fair Trade Certified products are sold at more than 45,000 outlets in the U.S. Biggest sellers: coffee and chocolate. Tea, sugar, rice, bananas, flowers, honey, and wine are also popular.

More than 400 college campuses now serve Fair Trade Certified coffee.

WHAT YOU CAN DO

Your Partner: TransFair USA, the only independent, third-party certifier of Fair Trade Certified products in the U.S. They guarantee Fair Trade transactions. See *FairTradeCertified.org*

Your Goal: Support small-scale farmers in developing countries by building consumer demand for Fair Trade products.

START SIMPLE

• **Look for the Fair Trade Certified label** when you buy coffee, tea, chocolate, sugar, vanilla, bananas, and rice. And request Fair Trade coffee at cafés. Going to Starbucks? Their Fair Trade Certified coffee is called Cafe Estima—ask for some. McDonald's and Dunkin' Donuts offer Fair Trade Certified coffee, too. Ask for it.

• **Download a copy of Co-op America's *Guide to Fair Trade*.** It's got all the details: *coopamerica.org/PDF/GuideFairTrade.pdf*

STEPS FOR SUCCESS

Step 1. Adopt a supermarket with friends, co-workers, or members of your congregation. Speak to the store manager on a regular basis and ask for more Fair Trade products. Plus, each time you shop, fill out customer comment cards asking for them. (P.S. Don't forget to shop at stores that *do* carry Fair Trade products.)

Step 2. Bring beliefs and practice together. Adopt Fair Trade Certified coffee, cocoa, tea, and sugar at your place of worship. There are already thousands of faith groups serving Fair Trade refreshments at congregation events.

Step 3. Have a party. Host a house party/product tasting with Fair Trade Certified products. (It doesn't have to be in a house.) Turn your friends and neighbors on to Equal Exchange Roasted Almonds or Dagoba Chocolate while you educate them about Fair Trade. Good ideas at *transfairusa.org/content/support/HP.php*

Step 4. Make Your Town a "Fair Trade Town." Modeled after Europe's successful Fair Trade Towns movement, TransFair USA's campaign helps you organize locally to make Fair Trade products the standard in the U.S.—one community at a time. For more info, visit *www.transfairusa.org/content/support/Fair_trade-resolutions.php#fttowns*

For more resources: *50simplethings.com/fairtrade*

22. CLEAN UP
THE MINES

Mining the gold that goes into the average wedding band (1/3 oz.) leaves behind an average of 20 tons of potentially toxic waste.

BACKGROUND. You might imagine that you have nothing to do with mining. But look around you: Everything you have that's metal—from silverware to cell phones, door hinges to soda cans—came from a mining operation.

Unfortunately, mining is one of the most polluting industries in the world, largely because few of us pay attention to it. As a result, companies that dig for gold, copper, bauxite, and other ores get away with decimating some of the most sensitive areas in the world. As concerned consumers, we can help change that.

DID YOU KNOW

• According to the mining industry's own reports to the EPA, they're our nation's leading toxic polluter—and have been for years. They're responsible for some 50% of all toxic emissions in America.

• Have you ever heard of the Summitville mine in Colorado? The Canadian company that ran it dumped toxic mining waste into the Alamosa River from 1984 until 1992, then declared bankruptcy. The area was left so polluted that the U.S. government had to declare it a toxic waste site. They're still trying to clean it up—using millions of taxpayer dollars.

• There are plenty of similar stories. The Zortman-Landusky gold mine in Montana spilled 50,000 gallons of poisonous cyanide solution into a community water supply, the Midnite uranium mine in Washington state contaminated local groundwater with radioactive materials (then refused to pay for cleanup), and so on.

• Mining pollution doesn't end when a mine closes. There are about 500,000 old, abandoned mines in the U.S....and many of them are *still* draining toxic acid into nearby streams. California alone has some 5,000 of these polluting sites.

A key chemical used to separate metal from ore in mining, especially gold, is cyanide—a poison.

WHAT YOU CAN DO

Your Partner: Earthworks. They've been dedicated to protecting communities and the environment from the destructive impacts of mining since 1988. Visit them at *earthworksaction.org*

Your Goal: Use consumer pressure and laws to reduce mining companies' environmental footprint and eliminate their worst practices. "Companies also need to hear directly from you that you want an alternative to 'dirty' metals," says Earthworks' Payal Sampat.

START SIMPLE

Save a can, save the Earth. One more reason to recycle cans (and cell phones, electronics, and other metal): It cuts down on mining waste. Every year, we throw out *1.5 million tons* of aluminum cans; 6 million tons of bauxite have to be mined just to replace them.

• Find a place—work, school, parks—where cans *aren't* recycled and put in a container. Make sure the cans get to a recycling center.

STEPS FOR SUCCESS

Step 1. Get on the Earthworks Action List. When there's an important mining issue that needs your help, Earthworks will let you know. Sign up at *earthworksaction.org*

Step 2. Bring the campaign to your community. When retailers and jewelers find out how destructive mining companies are, they often want to do something about it. The "No Dirty Gold" and Valentine's Day campaigns were designed by Earthworks to help retailers put pressure on the mining industry. Nearly 30 jewelry firms, including 7 of the 10 largest, have signed on already. Get your local jewelers and craftsmen on board. And sign on to the No Dirty Gold consumer petition. For info: *nodirtygold.org*

Step 3. Pressure corporations. Mining companies need to hear that their biggest clients—corporations—care about ethically mined metals. Earthworks will help you go beyond jewelry to influence other sectors of the economy, such as electronics and beverages. *Bonus: recyclemycellphone.org*

Step 4. Demand better laws. One reason mining companies were able to operate so irresponsibly is that U.S. mining laws governing their actions were written in the 1870s—when the West was still being settled! We need modern legislation to protect us, and to protect the environment. Visit *earthworksaction.org/us_program.cfm*

Mining and refining metals accounts for about 10% of the world's energy use.

23. DROP BY DROP

Water scarcity already affects every continent,
and 4 of every 10 people on Earth.

BACKGROUND. Experts predict that water is going to be "the oil of the 21st century." Actually, that may be an understatement: Fresh water isn't just essential to making our society run; we literally can't live without it.

All you have to do is look at what's happening with droughts, sinking water tables, and polluted waterways to understand that access to clean, fresh water is an increasingly serious—and potentially explosive—issue.

DID YOU KNOW

• There's plenty of water on the planet—an estimated 370 *billion billion* gallons. (Yes, you read that right.) But only about 2.5% of it is fresh water...and only about 3% of *that* is water we can actually use. The rest is stored in glaciers and snow cover.

• Think of it this way: If the world's entire water supply was 1 gallon, fresh water would make up 4 ounces, and readily accessible fresh water—our drinking, bathing, life-preserving water—would make up 2 *drops*.

• Population growth is creating water problems (see p. 112). So is pollution; it's estimated that we dump two million *tons* of waste into our precious rivers, lakes and streams every day. And development has turned forests and meadows into paved surfaces that can no longer absorb rain and snowmelt to replenish aquifers (see p. 40).

• Climate change is increasing the stress on our water supplies even more. U.N. scientists calculate that global warming will account for about 20% of the increase in water scarcity.

• People are already suffering from water shortages. According to the World Health Organization, "Today 1.1 billion people lack access to clean water and 2.4 billion lack access to proper sanitation." But it will get worse: "By the middle of the century," they say, "7 billion people may be faced with water scarcity."

Among 30 developed nations, Americans have the highest per-capita water use.

WHAT YOU CAN DO

Your Partner: American Rivers, a national organization that "protects and promotes our rivers as valuable assets that are vital to our health, safety, and quality of life." Meet them at *americanrivers.org*

Your Goal: Make using water wisely a focus of our society. "One of the challenges," explains American Rivers' Betsy Otto, "is that water-efficiency isn't sexy. People just don't perceive that we need to use water more wisely...until we run out."

START SIMPLE

Start saving. American Rivers' "3 simple ways to save water": 1) Fix leaks. A faucet leak of 60 drops per minute wastes 192 gallons a month; outdoor leaks from hose connections and irrigation systems can lose 50% of the water used. 2) Install low-flow showerheads and faucet aerators, which can cut flow by 50%. 3) Turn the faucet off when you brush your teeth or shave. You'll easily save over 5 gallons each time. More tips: *50simplethings.com/savewater*

STEPS FOR SUCCESS

Step 1. Start at home. How much water do you actually use? Look at your water bill and figure it out. How does that stack up against neighbors, or households similar to yours? Check your local water utility's Web site or use an online calculator. You can find links at *americanrivers.org/50simplethings* or *50simplethings.com/savewater*

Step 2. Watchdog your water utility. It's easier to save water if your utility helps. Check them out: Do they have conservation-based pricing that rewards wise water use? Do they offer free low-flow showerheads and other water-saving devices? Help move them in the right direction: Join with friends and neighbors, and persuade your utility to take some of these key steps. For advice: *americanrivers.org/50simplethings*

Step 3. Inspire a child. Work on bringing water-efficiency education to school. Our kids' lives depend on their understanding water issues. For links to a great water-saver curriculum and teaching tips: *americanrivers.org/50simplethings* or *50simplethings.com/savewater*

Step 4. Make it the law. Protecting our water supply is too important to leave to chance. We need to get local, state, and federal elected officials to *pass laws* to promote wise use of water. Get tips, techniques, and talking points at *americanrivers.org/50simplethings*

An average American uses 100+ gallons of water a day. An average Ethiopian uses 3 gallons a day.

24. BUILD IT GREEN

*Greensburg, KS, a small town wiped out by tornadoes in 2007,
became the first city in the U.S. to mandate all city buildings
be certified green. Their 10 new buildings will double
the number of green buildings in Kansas.*

BACKGROUND. When you think of the major causes of climate change, you probably think of car emissions, factories, or power plants. Most people don't realize the impact of buildings on global warming. They account for 36% of all greenhouse gas emissions in the U.S. They consume 71% of our electricity and 12% of our potable water—that's 15 trillion gallons a year.

Building codes have historically dealt with safety issues such as fire prevention, wiring, and structural standards, not with environmental issues. Even when people began putting solar panels on their roofs, there was no general definition of "green building" or how to tell if it really worked. Now there is: It's called LEED®, and it's literally helping us build a healthier world.

DID YOU KNOW

• LEED, which stands for Leadership in Energy and Environmental Design, is a green building rating system developed by the membership of the U.S. Green Building Council (USGBC) in 2000.

• LEED certification is like a nutrition label for buildings—it looks at water and energy usage, construction materials, where the materials came from, the quality of the indoor environment, and how it all can be used to reduce our collective footprint on the Earth.

• The annual U.S. market in green building products and services was over $7 billion in 2005, and over $12 billion in 2008.

• Examples of energy-reducing innovations used in green building: vegetation-covered roofs that collect rainwater and provide habitat, water-free toilets and low-flow faucets, and skylights for increased natural lighting. Unique design materials include insulation made from recycled blue jeans, recycled glass, and resin countertops.

• Green is not just for new buildings. In 2007, existing buildings became eligible for LEED certification as well.

Building-related construction and demolition account for 60% of non-industrial waste in the U.S.

WHAT YOU CAN DO

Your Partner: The U.S. Green Building Council (USGBC). Since its founding in 1993, the Council has grown to more than 13,000 member companies and organizations, and a network of 72 local chapters. USGBC offers a comprehensive family of LEED green building rating systems, education, and conferences.

You Goal: Help transform the building market by integrating and promoting green, "whole-building" design practices.

START SIMPLE

Go green. Download USGBC's *16 Great Ideas for a Greener Life* for ways to start making your home green right now. Our suggestion: Pick your five favorites and do them first. Get the file at *50simplethings.com/building*

STEPS FOR SUCCESS

Step 1. Change is good. Despite the success of green building in commercial and government building, the idea may still need a local push. Planning boards can be resistant to growth—even when it's green—and many zoning laws and building codes are outdated. Surveys show the #1 reason people don't build green is a lack of information.

• So educate your community about green building. Staff a table at community events with photos and case studies. Hold an open house for decision makers or arrange for a walk-through at a green home or a building under construction. Get resources from your regional USGBC chapter. Find them at *usgbc.org/chapters*

Step 2. Make it a career. Green building is a growing enterprise, and we realize this issue is going to have the most appeal to someone in the business—development, real estate, construction, architecture, trades, and so on. If you work in the building industry (or want to), or if you own or manage residential or commercial properties, you can join the effort to go green. Become a professional member of USGBC at the national level, and take advantage of their educational workshops, conferences, and other resources. More important, join your local or regional USGBC chapter and work to promote LEED certification for your projects and all others in your area.

Central air conditioners use 98% more energy than ceiling fans do.

25. BORN TO BE WILD

*An estimated 6,000 acres of open space are
lost each day, a rate of 4 acres per minute.*

BACKGROUND. Sometimes it seems as if America is being
taken over by big-box stores, fast food chains, new malls, and
developments.

If this bothers you, you might find it comforting to know that
America also has the opposite: huge tracts of legally protected natural space, where absolutely nothing can be built, not even roads.
These pristine pieces of land, officially called Wilderness Areas, are
exactly the same as they were 200 years ago—and they'll be the
same 200 years from now. The problem is, America has set aside
just a fraction of the wilderness that's needed to buffer the effects of
global warming, sprawl, and our burgeoning population. But you
can become part of the 21st-century wilderness movement.

DID YOU KNOW

• Wilderness Areas (which must be on federal public lands) are
important for many reasons. They protect watersheds, wildlife
habitat, and biodiversity. They improve air quality. They provide
outdoor recreation: hiking, hunting, camping, and other activities.

• There are 107 million acres of Wilderness now, and about 200
million acres more that experts believe should be saved. But if we
don't designate these areas as Wilderness quickly, we can lose them
forever. Why? As soon as roads or mining operations are built, the
areas may no longer legally qualify. So anti-environmentalists are
trying to develop areas before they can become Wilderness.

• In 2003, for example, the Interior Department announced it was
allowing development on millions of acres of Utah land that had
been proposed for wilderness designation in America's Red Rock
Wilderness Act, which was pending before Congress.

• One Utah activist warns: "[They're] pushing hard to scar these
lands with roads and oil fields, and take away from Congress its
prerogative to preserve these special areas for the benefit of future
generations." We have to push back now.

Less than 5% of the U.S. is protected as Wilderness.

WHAT YOU CAN DO

Your Partner: The Wilderness Society (TWS) has been protecting America's Wilderness since 1935. They wrote the book on wilderness protection and will work with you every step of the way to help you make a difference in a Wilderness campaign. Meet them at *wilderness.org*

Your Goal: Preserve America's threatened wilderness by helping to turn it into Wilderness Areas.

START SIMPLE

Adopt a Wilderness. There are always Wilderness proposals pending in Congress. Find out about current campaigns at *wilderness.org* Sign up for WildAlert, a weekly update on actions you can take.

• If there isn't one in your region to support, just pick one that sounds interesting. Then join the coalition that's spearheading the campaign; they'll be glad to have your help.

STEPS FOR SUCCESS

Step 1. Choose an area you think could be Wilderness. The campaign starts with local volunteers. First they do "ground truthing"—visiting the area, identifying boundaries and key resources, proving there are no roads. Volunteers create a Citizen's Wilderness Inventory—a detailed map and catalog of land and resources. You can see an example at *50simplethings.com/wilderness*

Step 2. Create a citizen's proposal. Work with TWS and a local partner group (such as the Arizona Wilderness Coalition) to craft the document that highlights the resources to be protected.

Step 3. Educate people and build local support. Enlist business owners, local officials (e.g., county commssioners, city council), and average citizens who'll write letters of support. Only Congress can designate Wilderness, and it's the really strong compaigns with lots of local support that get congressional support. So it's important that locals are well organized and speak with one voice.

Step 4. Take your proposal to your congressperson and get him or her to introduce it. TWS will work with you and your coalition to pass the federal legislation that protects your Wilderness Area. Think of it as a long-term project. The campaign might take years, but the area will be protected forever, thanks to your efforts.

175 of the 261 types of ecosystems found in the U.S. are protected in Wilderness Areas.

26. DOWN THE DRAIN

According to the EPA, as many as 75,000 sewage spills each year spew out 1.26 trillion gallons of untreated waste—a toxic cocktail of bacteria, viruses, chemicals, pharmaceuticals, hormones, and antibiotics.

BACKGROUND. We're taking one of our most basic necessities—clean water—for granted. What happens to the toxic chemicals that your neighbor uses to unclog the sink? When you flush the toilet...where does the waste go? You know it runs through pipes—but what are they made of? And who's taking care of them? And why do we have tens of thousands of beach closings from sewage overflows every year?

Unfortunately, in many communities the solution to pollution has been dilution. But dumping sewage into clean water does *not* make the sewage clean; it makes the water polluted. It's time to make *clean water infrastructure* a priority.

DID YOU KNOW

• In the U.S., the network of sewage water pipes extends more than 640,000 miles—four times the length of the national highway system. More than 16,000 sanitation facilities operate 24 hours a day, each one processing an average of 68 million gallons of water.

• According to the EPA, nearly 25% of the nation's water pipes are already "poor, very poor or elapsed." And they expect that number to jump to 45% by 2020.

• Each year up to 3.5 million illnesses are caused by swimming in water contaminated by sewage, and an additional 500,000 are caused by drinking contaminated water.

• Sewage pollution also affects aquatic wildlife. Scientists speculate that aberrations like the hermaphroditic fish found in the Potomac River and frogs that can't reproduce are the result of things we're putting in the water.

• Thirty-five years after the Clean Water Act, federal funding for clean water has fallen to the lowest level in its history. The EPA estimates that our water infrastructure budget is short by a whopping *$22 billion a year.*

The experts' advice: Try to avoid antimicrobial soaps. Like mercury, they build up in the...

WHAT YOU CAN DO

Your Partner: Food & Water Watch (FWW), a consumer organization working to ensure clean water and safe food in America and around the world. Get to know them at *foodandwaterwatch.org/water*

Your Goal: Cleaner drinking and recreational water in your community for today and for future generations. "Clean drinking water is one of the things that distinguishes our society," says FWW executive director Wenonah Hauter. "If we don't get control of our crumbling infrastructure, we're going to time-travel back 200 years when people got dysentery and cholera from their water."

START SIMPLE

Be kind to your pipes. Chlorine cleansers don't biodegrade. Use alternatives like oxygen- or hydrogen-peroxide-based bleach. Instead of drain cleaners which can damage water pipes, some people add one cup baking soda and one cup vinegar to a pot of boiling water (it will fizz) and pour it down the drain; they say it dissolves most clogs.

STEPS FOR SUCCESS

Step 1. Break the bottled-water habit. Bottled water is almost completely untested by the FDA and often has contaminants. It's important to build political support for funding public drinking water and sewage systems. If you have a problem with local drinking water, it's better to get a filter for your home. Take the pledge and find more information at *takebackthetap.org*

Step 2. Give your water a checkup. Take a look at the water quality report for your drinking water system. The EPA requires water systems serving more than a million residents to do at least 300 tests per month. Your water utility is required to provide this information. Call them or check their Web site.

Step 3. Get active in FWW's Take Back the Tap campaign. Advocate for a Federal Clean Water Trust Fund to help meet the $500 billion clean water infrastructure funding gap. We need a dedicated source of funding for constant improvement. Your activism can be as simple as signing a petition to Congress at *takebackthetap.org* or joining a local grassroots group to help spread the word. Contact FWW for help, or find links to additional grassroots organizations at *50simplethings.com/water*

...environment, and may contribute to the growth of antibiotic-resistant bacteria.

27. FOR THE BIRDS

Populations of some of America's best-known and beloved birds have plummeted over the past 40 years—some by as much as 80% percent.

BACKGROUND. Many of us take the birds in our backyards for granted, because it seems there are so many of them around. But according to the National Audubon Society, even many of our common birds are in trouble, while many rarer species are in danger of disappearing entirely.

Can you imagine a world without birds? It's hard to comprehend. But it isn't inevitable—this is a man-made crisis, and there are things we can do about it.

DID YOU KNOW

• About 45% of the bird species in North America are threatened or have declined in numbers over the past 40 years. These include meadowlarks, warblers, and thrushes.

• The primary reason: destruction of habitat. The forests, fields, and wetlands these birds need to survive are being lost to agriculture and development. Some birds literally have no place to live and breed.

• This isn't true only in North America. Many of our birds are migratory—they winter in specific tropical rainforests, which are being cut down to raise cattle and to grow soybeans and corn.

• Many birds are beautiful, but losing them isn't just a matter of aesthetics. Birds protect our food supply by eating billions of insects that damage food crops. And they're important natural gardeners: Look outside your window; many of the native trees and shrubs you see were "planted" by birds that dropped the plants' seeds there.

• Birds are also considered "environmental indicators"—they're sensitive to problems with the ecosystem. Experts say that when conditions on Earth are too fragile to support birds, then every living thing—including humans—is in trouble. So by helping birds, we're also helping ourselves.

Coffee drinker's alert: Buy "shade-grown" coffee. It's grown in the shade of rainforest trees...

WHAT YOU CAN DO

Your Partner: The National Audubon Society. They've been a part of every major effort to protect birds for over 100 years. Get to know them at *audubon.org*

Your Goal: Save and create bird habitat. "Start small, and take it as far as you want," says Audubon scientist Rob Fergus. "You'll get to know your yard, neighborhood and local habitats. You'll directly shape the world in a way that's better for you and for birds."

START SIMPLE

Invite birds into your yard. Put up a bird feeder—not only to provide birds with food, but also to make them a part of your life. Watching birds every day will make them something you enjoy and care about, not just an abstract idea of something to "save." Find out how at *audubonathome.org/bird_feeding*

STEPS FOR SUCCESS

Step 1. Learn about the birds in your area. If you're a beginner, get the basics from Audubon. Or pick out a few local threatened birds you want to help at *audubonathome.org/birdstohelp*. You can also join the Great Backyard Bird Count: Get to know your neighborhood's birds while you help experts keep track of bird populations in your area. For more info, see *birdcount.org*

Step 2. Plant a bird-friendly yard. Millions of people are creating habitat where they live. Your yard, your patio, even your windowsill are all habitat to birds. Get started at *audubonathome.org*

Step 3. Expand it to your neighborhood or community. Spread the habitat! Recruit sympathetic neighbors and develop a plan for your neighborhood or park. Download Audubon's neighborhood organizing guide at *audubonathome.org/neighborhood*

Step 4. Adopt an Important Bird Area (IBA). This is the "front line" of bird protection. Audubon is spearheading an effort to identify the most important locations for birds in each state, and focus conservation efforts on those sites. Find the IBAs in your area; get info on starting or joining an IBA group, at *audubon.org/bird/iba*

For more resources: *50simplethings.com/birds*

...so you're giving growers an incentive to preserve rainforest habitat. See *shadecoffee.org*

28. TEACH YOUR CHILDREN

The CDC reports that U.S. children ages 6–11 have more lawn-pesticide residues in their bodies than all other age groups.

BACKGROUND. About 20% of the U.S. population spends their days inside elementary and secondary schools. You probably assume these schools are safe for our kids (and their teachers), but it turns out that more than half of them have serious air quality problems. The sources of these problems—pesticides, asbestos, cleaning products, lead from paint, and so on—are not only affecting people's health, they're also polluting the planet.

Green Flag Schools, a free school program created by the Center for Health, Environment and Justice (founded by Lois Gibbs, famous for cleaning up Love Canal in the 1970s) teaches kids how to make their schools environmentally healthier places. It focuses on reducing pesticides and toxic cleaners, improving indoor air quality, and using nontoxic paint. Plus, it teaches students to save energy and make money with a recycling business. "There are many school programs that teach young people about the environment," says Gibbs. "The Green Flag Program is the first that [also] teaches them how the environment affects their health."

DID YOU KNOW

• According to the EPA, more than 53 million kids—plus 6 million adults—spend at least 6 hours a day, 180 days a year, in school.

• Children's bodies keep developing until adolescence; their systems are both more vulnerable to environmental toxins and less able to recover from their damage.

• There are more than 120,000 U.S. school buildings, many old and in poor condition. Kids in over half of them are exposed to indoor pollutants like mold, mildew, fumes from cleaners, and asbestos. These toxins have been linked to an increase over the last 20 years in childhood illnesses like asthma, leukemia, and autism.

The EPA offers an "Indoor Air Quality Tools for Schools Action Kit." It's free to schools: *epa.gov/iaq*

WHAT YOU CAN DO

Your Partner: The Green Flag Program for Environmental Leadership, which is a project coordinated by the Center for Health, Environment and Justice (CHEJ). Through the Green Flag Program, CHEJ extends its long tradition of community organizing assistance to individual schools. Meet CHEJ: *chej.org*

Your Goal: Create an effective environmental program in local schools, and educate your community about advocacy and problem-solving in the process.

START SIMPLE

Complete the Back-to-School Environmental Checklist at *50simplethings.com/greenflag*. Contact your local school and arrange a walk-through with this 10-question checklist. Use the results to decide what issues need the most attention first.

STEPS FOR SUCCESS

Step 1. Get started. Anyone can start a Green Flag Program—parents, students, faculty, or community advocates. And it's free! Visit *GreenFlagSchools.org* and choose K–8, 9–12, or Green Flag/Green Faith Schools. Download materials, consult CHEJ, then host a meeting to put your Green Flag Team together.

Step 2. There's no I in team. Green Flag Team members will include adult school and community members, but the Team will mostly be made up of students, and they'll conduct the program. CHEJ will show you how to obtain the proper permissions from school and local authorities.

Step 3. Plant a Green Flag. The team will choose one of four project areas: *Reduce, Reuse, Recycle; Integrated Pest Management* (IPM); *Indoor Air Quality* (IAQ); or *Non-Toxic Products.* Each has a three-level plan: *Level 1:* Assess your school, form a Green Flag Team; *Level 2:* Discover and share information; *Level 3:* Create or improve a policy. Details here: *greenflagschools.org/levels.htm*

Step 4. Celebrate. During and after the project, kids on the team will get awards like badges, press articles, and the *Green Flag Award for Environmental Leadership.* You can hold your own community event to highlight their accomplishments.

Between 1993 and 1996, there were 2,300 pesticide poisonings in schools.

29. MAKE NEW DIRT

It takes hundreds of years to grow just one inch of topsoil,
but we lose 25 billion tons of it worldwide each year.

BACKGROUND. Can you imagine your PB&J sandwich or the eggshell from this morning's breakfast turning into dirt? For that matter, can you look at dirt and imagine it was once an apple core?

If it seems like there ought to be something you can do with your leftover food besides throwing it away, you've come to the right page—the one about composting. You probably already know that composting is a way of recycling organic material, letting it become soil as a part of nature's cycle...instead of dumping it in a landfill, where it's actually *prevented* from enriching the Earth.

But composting is a way to help prevent global warming...and to create one of the most precious things on Earth: topsoil, the nutrient-rich dirt that makes all plant life possible.

DID YOU KNOW

• Because of erosion, topsoil is being lost from our fields 10 times faster than it can be replaced. According to geologist David Montgomery, author of the book *Dirt*, we're losing about 1% per year. "Globally," he says, "we're running out of dirt."

• Composting makes new dirt, but most Americans don't compost. In fact, organic materials like paper products, food scraps and yard trimmings make up 50% of the waste stream that goes to landfills.

• When biodegradable material like food and yard waste is trapped in a landfill with no air, it doesn't create topsoil—it ferments and creates a powerful greenhouse gas called methane. According to the International Panel on Climate Change, methane is 25 times "more effective than CO_2 in causing climate warming." And landfills generate more methane than any other human source.

• Community composting is a new kind of recycling program that keeps organic material out of landfills. A pilot program in San Francisco has been wildly successful; they expect to reach 75% recycling by 2010. They already sell the compost they make.

Bad compost! It takes approximately 350 years for an aluminum can to decompose.

WHAT YOU CAN DO

Your Partner: Eco-Cycle, the largest nonprofit community recycler in the country. Get acquainted at *ecocycle.org*. Eco-Cycle will be your bridge to the national community composting network.

Your Goal: To get biodegradable materials out of landfills, and make nutrient-rich new dirt for your community.

START SIMPLE

• *To compost or not to compost?* Composting sounds like a no-brainer. Who doesn't want to—literally—save the Earth and divert 50% of the waste stream? But it does take some work. Learn more about it so you can decide how (or whether) to make it part of your life. Check the links at *50simplethings.com/compost*

• *Take a composting class.* Get some inspiration and a firsthand look at how to do it. Your local recycler or co-op may give classes, or check with your county extension services for advice.

STEPS FOR SUCCESS

Step 1. Compost at home. If you take the plunge, make sure you have expert guidance and the right equipment—a composting bin, mulching mower, or worms (yes, worms). Your county extension service can get you started, or check online resources: *ecocyle.org*, *earth911.org*, and *50simplethings.com/compost*

Step 2. Compost at school. Kids love dirt; let them make some. There are plenty of resources online for teachers, and lots of stories about successful organic recycling programs at school. We've researched it and have the links at *50simplethings.com/compost*

Step 3. Compost with your community. Community composting is the perfect solution for residents, businesses, and schools alike. Recyclers supply three color-coded curbside bins: one for conventional recycling, one for food and garden waste, and a very small one for nonrecyclable refuse. For a project like this, forming alliances is very important. So start discussing it with a small group of interested neighbors—your garden club, environmental groups, etc. Eco-Cycle is the expert on creating community composting programs, so contact them at *ecocycle.org*. For more info, check out *50simplethings.com/compost*

A compost pile can reach 150 degrees Fahrenheit.

30. CONSIDER THE ALTERNATIVES

Technically, there's enough wind power in the U.S.
to meet our electricity needs four times over.

B ACKGROUND. Have you ever watched a pinwheel spinning in the breeze? That's "alternative" energy. Felt the heat of the sun on your face? *That's* alternative energy. Watched a stream carry a stick out of sight? That's alternative energy, too.

It's funny that people call it *alternative* energy when it's really about as basic as you can get. And it's amazing we haven't done more with it. After all, it's natural, clean, safe, accessible, and virtually unlimited.

So let's call it "clean, renewable energy" instead. It's time to stop thinking of it as an alternative…and start thinking of it as a necessity.

DID YOU KNOW

• Our current power system is unsustainable. More than 70% of our electricity comes from fossil fuels (coal, gas, oil), which are finite and heavy polluters. Coal-fired power plants, which produce about 50% of our electricity, are the single greatest contributor to greenhouse gas emissions in the U.S.

• The use of wind power is increasing 25–40% per year worldwide. Denmark currently gets a full 25% of its electricity from wind.

• Geothermal energy uses heat from Earth's crust to produce electricity. In Iceland, geothermal-heated soil is used to grow *bananas*.

• Hydrokinetics is a developing technology that produces electricity using the natural kinetic (moving) energy in rivers, waves, and ocean currents.

• Bio-energy comes from organic plant matter—from crops to native plants to agricultural waste—and can be burned directly or "gasified" to produce power.

• Solar power (see p. 24): The sunlight that shines on the Earth in one hour could meet world energy demand for an entire year.

The cost of producing wind power has fallen by as much as 90 percent since 1980.

WHAT YOU CAN DO

Your Partner: The Union of Concerned Scientists (UCS), a leading science-based nonprofit working for a healthy environment and a safer world. UCS is a leading voice for renewable energy policy and technology. Check them out: *ucsusa.org*

Your Goals: Learn about clean, renweable energy, get other people excited about it, and work in your community to support it.

START SIMPLE

• *Buy green energy.* Many utilities already offer clean energy at a small premium. Pacific Power, for instance, calls their program "Blue Sky"; Tennessee Valley Authority's is "Green Switch." You may be able to sign up simply by checking a box on your utility bill. Contact your power company to find out more.

• Even if your utility doesn't offer green power, you can still purchase it in the form of renewable energy certificates (RECs). Most homes can offset their electricity use with RECs for as little as $10 to $20 a month. See *50simplethings.com/renewables*

STEPS FOR SUCCESS

Step 1. Learn more about renewable energy. Since you're reading this book, chances are that learning about renewable energy will be pretty exciting to you. You may be amazed to find out how many renewable energy projects are already under way in your area. Find them on the U.S. Department of Energy's Web site at *http://tonto.eia.doe.gov/state*

Step 2. Be demanding. If your utility doesn't offer renewable energy, ask for it. If your town, city, or state is considering a renewable energy facility, attend public meetings and show your support. It makes a difference.

Step 3. Support a renewable electricity standard. Another way to get utilities to commit to renewables: *make* them do it. Pass a *renewable electricity standard* in your state and in the federal government. This law gives utilities a deadline by which they must increase renewable energy to a certain level. In Oregon, for example, it's 20% of electricity by 2025. At the moment, only 25 states have this law. And there's no national standard. Get updates and learn how to make your voice heard by signing up for the UCS Action Network at *ucsaction.org*

Largest consumers of coal in the world: 1) China, 2) the U.S.

31. IT'S A
PLASTIC WORLD

Americans throw away 2.5 million plastic bottles every hour.

BACKGROUND. Take a look around your home and try counting the things made of plastic. It could take you all day—plastic is used in practically everything we buy.

Plastic is a miracle material, but it has serious environmental drawbacks: It's made of nonrenewable fossil fuels; manufacturing it creates pollution and toxic waste; and it's not biodegradable. On top of that, we throw most of it away.

Clearly, we need to cut down on "disposable" plastic. But we also need to change the way plastic is made, and the materials it's made from. That's why people are working on *bioplastic*, which is biodegradable, compostable, made from renewable and sustainably harvested materials...and works as well as "normal" plastic.

DID YOU KNOW

• "Bioplastic"—plastic made from plants instead of fossil fuels—isn't a new concept. The very first synthesized plastic, created in 1845, was a cotton-derived material called *celluloid*. By the 1870s, it was used widely for dental plates, combs, and toothbrushes.

• Today, manufacturers make an estimated 200 million tons of plastic a year. Less than 3.5% is recycled...which means that every year, we add 193 million tons of plastic to the world—permanently. About 25% of our landfill space is taken up by plastics.

• Not all of the plastic winds up in landfills. In a part of the Pacific Ocean known informally as the *Pacific Garbage Vortex*, there's a bigger mass of plastic than there is of plankton.

• Consumer alert: Bioplastic isn't automatically good for the environment just because it's made from plants. Some bioplastic is made from GE crops (see p. 100), some doesn't biodegrade, and some has toxic additives; some interferes with recycling and some contains engineered nanoparticles, particles so small they can pass in and out of our cells. It takes some research to tell the difference.

It's estimated that every three years, the amount of plastic in the ocean doubles.

WHAT YOU CAN DO

Your Partner: The Institute for Local Self-Reliance has been "working to strengthen communities through the smart utilization of local resources" since 1974. One current focus: creating sustainable bioplastics. Get acquainted: *ilsr.org* and *sustainableplastics.org*

Your Goal: Cut down on disposable plastics and help make sustainable bioplastics a viable alternative.

START SIMPLE

• "Choose durability over single-use disposable plastics," says Brenda Platt of ILSR. "Can you use a travel mug instead of a polystyrene cup, reusable shopping bags instead of plastic or paper? After we reduce consumption, we can move on to choosing better materials."

• So…pick a few disposable plastic items you use, and dispose of them—permanently. For suggestions: *50simplethings.com/plastic*

STEPS FOR SUCCESS

Step 1. Become a plastics expert. Read about the different polymers in the *Plastics Guide* you'll find at *50simplethings.com/plastic*. And read about bioplastics at ILSR's *sustainableplastics.org*

Step 2. Spread the word. Let's assume the world isn't going to quit using disposable plastic items overnight. One way to get the word out is to substitute new bioplastics for the worst of the fossil-fuel-based stuff. Get your PTA or congregation to switch to potato- or corn-based compostable forks, spoons, and cups to replace polystyrene.

Step 3. Lobby businesses and institutions to switch their takeout containers and utensils to compostable natural fibers. Start with sympathetic people: natural food stores, green restaurants, coffee shops. Then target local schools, public events, and festivals.

Step 4. Start composting. Compostable plastics don't make sense unless they're composted. When they're landfilled, they give off methane, a powerful greenhouse gas. See p. 76 for more info.

Step 5. For the ambitious: Get the most harmful and wasteful plastics banned locally. Oakland, San Francisco, and two dozen other cities are already banning polystyrene takeout containers and encouraging the use of alternatives. Push your local government to do the same. Get more info at *50simplethings.com/plastic*

According to the *Wall Street Journal*: Americans use 100 billion plastic shopping bags each year.

32. HOME ON THE REFUGE

Scientists have predicted polar bears could disappear from Alaska within 50 years, which makes their protected habitat in the Arctic National Wildlife Refuge a crucial lifeline.

BACKGROUND. There's no feel-good way to spin it: Animals and plants need places to live, and modern society is destroying those places. Unchecked development, pollution, and global warming are all degrading or eliminating the habitat that animals—from grizzly bears to songbirds—need to survive. And once their habitat is gone, the wildlife will be, too.

We may not be able to stop that from happening on a large scale, but we can do a little at a time…which is why National Wildlife Refuges are so important. These legally protected areas, dedicated exclusively to wildlife preservation, are among the only places where volunteers can focus on improving—rather than simply trying to save—habitat for threatened plants and animals.

DID YOU KNOW

• Many Americans aren't aware that the National Wildlife Refuge System exists, but it includes almost 100 million acres—which makes it larger than our National Park System. There are over 547 wildlife refuges scattered in all 50 states and U.S. territories.

• National wildlife refuges preserve nearly all of America's habitats, from Arctic tundra to tropical islands, from arid desert to lush coastal wetlands.

• About 25% of refuges are considered "urban"—i.e., they're near metropolitan areas with populations of 50,000 or more. Annually, nearly 40 million people visit refuges to hunt, fish, birdwatch, and take advantage of educational opportunities for children.

• Animals already have to deal with rampant development like urban sprawl, walls, roads, and dams. With climate change accelerating at an unprecedented rate, it's becoming even harder for them to survive. Refuges provide safe havens where many threats are eliminated and wildlife has a chance to adapt and thrive.

The first National Wildlife Refuge was created by President Teddy Roosevelt in 1903. The reason?…

WHAT YOU CAN DO

Your Partner: Defenders of Wildlife. Since 1947, they've worked to protect and restore America's native wildlife and its habitat, resolve conflicts, and educate and mobilize the public. Get to know them at *defenders.org*

Your Goal: Support the National Wildlife Refuge System. "If you love wildlife, you'll find this an easy, rewarding task," says Defenders President, Rodger Schlickeisen.

START SIMPLE

Buy a stamp, create a refuge. Buy a Federal Duck Stamp at the post office for $15. Not only do Duck Stamps provide free admission to all refuges, but 98¢ of every dollar goes directly to buy or lease wetlands for protection in the Refuge System. The stamp is artistic, collectible, and good for a year: *defenders.org/duckstamps*

STEPS FOR SUCCESS

Step 1. Visit a National Wildlife Refuge. There's one about an hour's drive from every major urban area in the country and it's a great family outing. Most have free admission. If not, use your Duck Stamp to get in. All you need to know is at *defenders.org/refuge*

Step 2. Volunteer on a Refuge (*fws.gov/volunteers*). About 20% of all work on refuges is done by volunteers. Help is needed in every area, including battling invasive species, planting native grasses and trees, guiding school tours, and working in the Visitor's Center. Volunteer one Saturday a year, or go once a week—it's your choice.

• Defenders of Wildlife offers great opportunities for volunteer work on refuges through their Wildlife Volunteer Corps. Visit *defenders.org/wildlifevolunteercorps* for upcoming events.

• You can also join a Refuge Friends Group. Find your nearest wildlife refuge at *defenders.org/refuge* and contact the refuge directly.

Step 3. Invite your congressional representative to visit a refuge with you. Nothing will make a bigger difference to a congressperson's understanding and support than seeing a refuge with a group of dedicated volunteers. Call their local and D.C. offices. If they're reluctant or unavailable, invite staffers first. This will take persistence and time; contact them regularly until they accept your invitation. Then invite the media! See *defenders.org/advocacy*

...To protect endangered birds from hunters who wanted their colorful plumage for hats.

33. ALL ABOARD!

*The Acela Express is Amtrak's model for success. It carries
54% of the 10,000 commuters who travel the corridor
between Washington D.C., NYC, and Boston every day.*

BACKGROUND. Have you taken a trip by train lately?
Probably not. Although railroads have long been a major
part of passenger transportation systems all over the world, in
the U.S. they were generally considered irrelevant—until recently,
when Amtrak's ridership began setting records each year.

The recent surge in train ridership is easy to understand: Modern trains are one of the most effective, economical, and ecologically sound ways of moving large numbers of people from one place to another. And with today's technology, trains operate both quickly and comfortably. So why *isn't* our government investing in a modern railway system?

DID YOU KNOW

• On a per-passenger-mile basis, trains are *the* most energy-efficient form of transportation, using 18% less energy than cars and planes. And new high-speed trains also cut down travel time. In France, for instance, one high-speed train travels at 200 mph.

• Trains reduce air pollution and greenhouse gas emissions. A train carrying 200 people 300 miles, for example, emits 22 fewer tons of CO_2 than 200 cars, and 15 fewer tons than two jets.

• Other industrialized nations have tracks dedicated to passenger trains. But in the U.S., the tracks are privately owned by freight companies. So while passenger trains are *supposed* to get priority, they often don't. The inevitable result: A third of Amtrak trains are late. Despite that, more people are riding Amtrak trains than ever—26 million in the year ended September 30, 2007.

• Studies show that even more people would ride if there were more trains and they were scheduled conveniently. In 2006, rail advocates persuaded the Illinois legislature to double its appropriation for Amtrak service, and Amtrak was able to add more trains to its schedule. The result: Ridership exploded, increasing by 80%.

Amtrak officially began service on May 1, 1971, with 25 employees. Today, they have over 24,000.

• But the U.S. *still* refuses to invest in passenger trains. In 2006, Amtrak, which is federally funded, was allocated only $1.3 billion. Sounds like a lot? Privately owned airlines got $16 billion of taxpayer money…and highways received $40 billion. According to Worldwatch Institute, when additional taxpayer costs are figured in, the *actual* subsidy for highways is $300 billion a year, or $2,400 per automobile.

WHAT YOU CAN DO

Your Partner: The Environmental Law and Policy Center (ELPC), a Midwest environmental advocacy organization focused on improving the environment and the economy. Join ELPC's efforts to expand high-speed rail in the U.S. Visit *elpc.org*, where you can send letters to legislators in support of better train service.

Your Goal: Help create a healthy railway system in the U.S.

START SIMPLE

Get on board. Join the National Association of Rail Passengers (NARP), an organization for people who ride trains. Sign up at *narprail.org*. Get action alerts and the latest news about rail transportation. (And, frankly, where else could you find it?) Learn what people are doing to improve U.S. passenger trains.

STEPS FOR SUCCESS

Step 1. Join a local (or regional) pro-rail group. There are more than 30 state chapters of NARP, plus a number of independent groups—and state chapters of groups like the Sierra Club are often involved with rail. Find out what's going on and how you can help. Your ideas could have a real impact on making this happen. Advice and info: *narprail.org* and *50simplethings.com/rails*

Step 2. Ride the train whenever you can. Every ticket purchased is a vote for a better train system.

Step 3. Be an advocate. This issue is about getting state or federal governments to change their priorities. That means changing minds in government and on a community level. Tell your elected officials about high-speed trains. And if your state has train service or is considering it, contact state government officials to vote "Yes!" Tell them that railway dollars are well spent and should be increased. For more info: *elpc.org*

There are over 200,000 miles of railroad tracks in the U.S.

34. SAVE THE
WETLANDS

Wetlands cover only 4–6% of the Earth's surface yet their global functions are estimated to be worth over $200 trillion annually.

BACKGROUND. Does a wetland have to be wet? The answer seems obvious, but it's really a trick question, because wetlands *don't* have to be wet all the time. (Prairie potholes are wetlands that are dry for half the year.) The point is, most of us don't know much about wetlands—even though they're critical to our planet's environmental health. It's time to learn.

DID YOU KNOW
• Wetlands are generally defined as land that's periodically flooded—whether it's by tides, river flows, rain, or groundwater. Equally important: Those areas have to be wet often enough, and for a long enough time, to support vegetation that's adapted for life in saturated soil. Wetlands can be as ordinary as a streamside marsh in your town or as exotic as a mangrove swamp in the Florida Keys.

• Wetlands are found on every continent except Antarctica, and in every climate from the tropics to the frozen tundra. In fact, there's probably one in your community.

• Wetlands come in all shapes and sizes: Some are as small as a pool table, others are as large as the state of Delaware. Some wetlands are freshwater, some are salt. Some have trees, others have flowering plants and grasses.

• Between the 1950s and 1970s, more than a half-million acres of wetlands were being lost every year in the U.S. By the mid-1980s, the nation had lost over half of our original wetlands.

• Fortunately, the rate of net wetland loss has slowed over the last 30 years…but we're still losing over 80,000 acres annually. This is equivalent to losing a football field of wetlands every nine minutes.

• What are we losing? Wetlands are among the most productive ecosystems in the world; acre for acre, there's more life in a healthy

More than 95% of California's wetlands have been lost.

wetland than in almost any other habitat. All species of fresh water fish depend on wetlands, up to 50% of our bird species live or feed there, and 50% of our endangered species depend on them.

• Wetlands also play a vital part in improving water quality (they remove nutrients such as nitrogen and phosphorus from the water), preventing erosion, and controlling floods.

WHAT YOU CAN DO

Your Partner: Environmental Concern has been working with wetlands since 1972. They'll support your efforts with information and advice. Get acquainted at *wetland.org*

Your Goal: Protect, restore, or create wetlands in your community.

START SIMPLE

• *Go exploring.* Visit a local wetland—see what you're saving. (And take your kids!) It's easy to find one: Call a local environmental group, your parks department, or the Dept. of Natural Resources.

• *Make the field trip more fun.* Download the one-page Wetlands Walking Guide from *wetland.org/guide* and take it along. More tips: *wetland.org/vi*

STEPS FOR SUCCESS

Step 1. Adopt a wetland with a local community group. Learn about the functions and value of wetlands, and the types of plants and animals that inhabit the area where you live. Then get your hands wet—be a weekend warrior with your group, planting and cleaning up your wetland. Or check out a school ecology club; they might need your help with a wetland project. Celebrate World Wetlands Day on February 2, and Wetlands Month in May.

Step 2. Join a local watershed association. Become a member and help support efforts in your own backyard. Environmental Concern will help you find one.

Step 3. Volunteer to preserve and restore wetlands in your community or on your property, where former wetlands have been destroyed. There are fascinating opportunities wherever you live. Enviomental Concern will help you find the right kind of project. For more information, check out *wetland.org* or *50simplethings.com/wetlands*

"When the well is dry, we know the worth of water." —Poor Richard's Almanack

35. WE'VE GOT CHEMISTRY

In a 2005 study, babies were tested for 413 industrial compounds.
On average, the babies had 200 of them in their blood.

BACKGROUND. Modern chemistry has undeniably made life easier. We've been able to grow more food per acre...prevent fires...make perfect scrambled eggs with easy cleanup. In the last 60 years, we've seen more chemical innovations than any other time in the history of the world. Since 1945, more than 20,000 new chemicals have been created and put to use.

Many of the 80,000 chemicals currently on the market were introduced into consumer products so quickly that there was no time to adequately examine their impact on people or the planet. "By using all these chemicals before we know they're safe," says Bill Walker of the Environmental Working Group, "we're conducting a vast experiment on ourselves and the world." We can't take back chemicals already released into the environment, but we can stop producing the most damaging ones, ensure safe use of others, and reduce our exposure to harmful chemicals in everyday products.

DID YOU KNOW

• It can take decades to see a chemical's real impact. Freon was introduced in 1928 as the "miracle compound" that made modern refrigerators possible. It took about 45 years to realize that freon was putting all life on Earth in jeopardy by depleting the ozone layer.

• Once chemicals are in the environment, there's no way to control them. Perfluorochemicals (PFCs), used in Teflon cookware, turn up in the Arctic and in South Pacific bird sanctuaries. They've been detected in 76 of 98 animal species tested, in 14 countries.

• Chemicals are persistent. The insecticide Mirex, toxic to shellfish and humans, was banned after 16 years of use in the southeastern U.S. Three decades later, it's still found in the environment.

• Change is possible. When the EPA took lead out of gasoline, blood lead levels declined dramatically. Legislation like the Kid Safe Chemicals Act would accomplish a similar goal for hundreds of chemicals by drastically reducing exposures to child-safe levels.

The outdated *Toxic Substances Control Act* makes it almost impossible for the EPA...

WHAT YOU CAN DO

Your Partner: The Environmental Working Group (EWG). Founded in 1993, their mission is "to use the power of public information to protect public health and the environment." Visit them: *ewg.org*

Your Goals: Learn about your family's exposure to chemicals, take steps to protect yourselves, work for better regulation of chemicals. "Our job is getting you the information you need to make safe choices," says EWG's VP for research, Jane Houlihan.

START SIMPLE

Pollution solutions. Four suggestions from EWG: 1) Use cast-iron pans instead of nonstick; 2) Never microwave plastic; 3) Buy organic, or eat produce known to be low in pesticides (*foodnews.org*); 4) Filter your tap water for drinking and cooking. Get details and more tips at *ewg.org/solutions*

STEPS FOR SUCCESS

Step 1. Hidden chemicals. EWG suggests you pick one area to look at the chemicals in your life, such as the cosmetics and personal care items you use. On average, they say, men are exposed to some 85 chemicals in those items a day, and women are exposed to over 150. Go to *cosmeticsdatabase.com* to see how your products score.

Step 2. Keep looking. Examine other household products your family uses, including cleaners, pesticides, and herbicides. Replace the worst offenders with natural alternatives or other less-toxic products. And don't throw old products down the drain. Check with local authorities about local hazardous waste drop-off programs.

Step 3. Speak up. If you stop using a product because of the chemicals it contains, let the manufacturer know. Contact customer service via their Web site, send letters as well as e-mails to top management, and follow up with phone calls.

Step 4. Find the hidden toxins. Find out about products used in local schools, day cares, and other public places (see p. 74). Circulate a petition directed at school boards or other relevant authorities, asking them to switch to safer products.

Step 5. Pester policy makers. Encourage federal and state lawmakers to adopt new rules governing the labeling and use of toxic chemicals. Details about current campaigns: *ewg.org/action*

...to ban *known* toxic chemicals like asbestos—let alone those that *might* pose risks.

36. WHAT'S BUGGING YOU?

More than 500 insects and 270 species of weeds have now developed resistance to one or more pesticides.

BACKGROUND. When DDT was introduced in the 1930s, it was regarded as a miracle of modern chemistry. After thousands of years, humanity finally had a "safe and effective" silver bullet to keep pests at bay. But DDT turned out to be toxic to insects *and* other animals—including humans (it persists in our fatty tissue). By the 1950s, its heavy use had pushed eagles and other predatory birds to the edge of extinction in America. In the 1970s, it was banned widely, except for temporary malaria control.

The DDT story is a classic example of the problems created by unexpected side effects of many synthetic pesticides. They've become a health and ecological disaster, and as they're overused, weeds, insects, and bacteria build resistance to them. Fortunately, there are effective alternatives. As consumers, we should encourage farmers to use them, and learn to use them in our own homes.

DID YOU KNOW

• Integrated pest management (IPM) is the best way to prevent pest problems without routine spraying of herbicides, insecticides, rodenticides, or fungicides. It has a long track record of success.

• In 1986, for example, an insect called the brown plant hopper was destroying Indonesia's vital rice crop. Why? The hopper had become resistant to chemical pesticides introduced by the "Green Revolution"...while its natural predators, dragonflies and spiders, were killed by the pesticides. President Suharto declared IPM a national priority; since then, Indonesian farmers have used IPM to bring back not only rice, but also other crops—and IPM has spread across the developing world.

• Closer to home: In 2005, San Francisco successfully fought West Nile Virus with IPM. Instead of aerial spraying with toxic pesticides, they trained bicycle messengers to drop packets of a biological insecticide that acts as birth control for mosquitoes into the city's 20,000 storm drains.

The average American home contains 6.5 gallons of hazardous materials.

WHAT YOU CAN DO

Your Partner: Pesticide Action Network (PAN) North America combines science and community-led campaigns to force global phaseouts of highly hazardous pesticides. PAN connects pesticides to larger issues—including environmental health. See *panna.org*

Your Goal: Build awareness of the problems with pesticides and the IPM solution. Change the way you use them in your life, and make it an issue in your community.

START SIMPLE

Support the alternative. Buy organic when you can—not just food, but flowers, bedding, clothes, seeds, and nursery plants. It's healthier for people, wildlife, and the very future of agriculture.

STEPS FOR SUCCESS

Step 1. Learn more about pesticides and IPM. At *panna.org* you can find a huge pesticide database, a "Nonpesticide Advisor" for tips, a magazine, and weekly news e-mail. Sign up for their pesiticide action alerts. For other links: *50simplethings.com/ipm*

Step 2. Use IPM at home. Indoors or out, a home program can be effective: Lawns and gardens can be the largest source of pesticide runoff. Contact your local sanitation or water agency for help—they often have really good info, guides, or even IPM hotlines because polluted runoff is such a big problem for them. For example, check out *ourwaterourworld.org*. If you rent, get your property manager, landlord, or co-op on board with IPM.

Step 3. Bring IPM to your community. Work to get IPM used at local parks and schools (see p. 74). Talk to your neighborhood association about using IPM for lawn care; see *panna.org/resources/lawns*. Make IPM part of your business's EPP program (see p. 94).

Step 4. Fight drift! Does your state or county allow pesticide spraying to control mosquitoes or moths that affect crops? First task: Get them to notify the public about crop dusting or pesticide use in public buildings and schools. Next task: Get them to adopt IPM. Find a local campaign: *panna.org/drift/fight*

Step 5. Work for laws that protect people and the environment from chemical harm. Join PAN and other groups to tell the EPA and Congress to label all chemical ingredients and assess alternatives as well as risks: *action.panna.org/* or *50simplethings.com/ipm*

In June 2007, Connecticut enacted the first ban on lawn pesticides on all outdoor school grounds.

37. SAVE THE
LIVING RIVERS

*The 2,000-mile Rio Grande, our 5th-longest river, is
so overtapped that at times portions of it run dry.*

BACKGROUND. Rivers are symbols of America's heritage,
but the rivers we know today are hardly the ones our ances-
tors knew. For 150 years, we've obstructed river flows with
dams and concrete channels; we've dredged and straightened big
rivers like the Missouri or Tennessee so people could build on their
floodplains; we've destroyed the wetlands and wildlife that lived on
the banks or in the water.

Miles of waterways that existed 100 years ago are no more than
ghost rivers today...and the waterflows of our *active* rivers are being
depleted all over the country. We're placing too many demands on
them—they simply can't keep up. It may sound crazy, but we're
reaching the point where we have to worry about whether there's
enough water left in our rivers for basic environmental needs after
cities, industry, and agriculture get through with them. And if there
isn't, what are we going to do about it?

DID YOU KNOW

• According to the journal *Science*, studies show that as global
warming increases, snow packs in the West are declining, causing a
growing number of rivers to run dry in the spring.

• Already, the Colorado River rarely makes it to the sea. "With the
states of Colorado, Utah, Arizona, Nevada, and, most important,
California depending heavily on the Colorado's water," explains
one report, "the river is simply drained dry before it reaches the
Gulf of California. This excessive demand for water is destroying
the river's ecosystem, including its fisheries."

• In the East, severe drought—made worse by global warming—
threatens rivers. According to the National Climate Data Center,
federal officials have declared 43% of the contiguous U.S. to be in
"moderate to extreme drought." In the Southeast it's a "mega-
drought," and states are fighting over water.

The United States has over 250,000 rivers. That's 3.5 million miles of rivers.

- Atlanta, for example, gets 70% of its water from the Chatta-hoochee River system. With its population growing, and facing the "worst drought in 500 years," Georgia is trying to hoard water by cutting the river's flow into Alabama and Florida— which could damage Florida's seafood industry, wipe out at least one endangered species, and shut down a nuclear power plant.

WHAT YOU CAN DO

Your Partner: WildEarth Guardians protects and restores the wildlife, wild places, and wild rivers of the American West.

Your Goal: To help keep America's rivers *living* rivers, with as much of their water flow preserved as possible.

START SIMPLE

Use water wisely. 63% of the country gets drinking water from water systems that use surface water sources, and in many communities rivers are the primary surface source. Wasting water is wasting our rivers. See "Drop by Drop" on page 64.

STEPS FOR SUCCESS

Step 1. Join a Friends of the River group. If you live near a river, chances are there's a local group that supports it. Make them stronger by joining. Get more info at *50simplethings/river*

Step 2. Work to establish a Living River Fund. This program, established through local governments, collects money to ensure the healthy flow of your river by buying water rights to it. Consumers check a box on their water bill. In Santa Fe, Albuquerque, and other cities along the Rio Grande, more than 30,000 acre-feet of water will be "put back" in the river for environmental needs like habitat protection. Download the WildEarth Guardians "road map," with logos, sample bills, and how-to instructions: *wildearthguardians.org/50simplethings*

Step 3. Advocate pro-river policies. Your state or community can require that a river's ecological needs be considered first. The Washington County Conservation District, in southern Utah, puts as much as 10% of water conserved "back into the river" as a reserve. States like Wash. and Mass. have laws guaranteeing enough water is left in rivers to meet their ecological needs. More info: *americanrivers.org* or *50simplethings.com/rivers*

The 3 longest rivers in the U.S.: 1) The Missouri 2) The Mississippi 3) The Yukon

38. GREENER BY THE DOZEN

"Even if you're against the idea that climate change is important, why in God's name would you be against saving money?" —Wal-Mart CEO Lee Scott

BACKGROUND. Wouldn't it be nice if government, schools, and big business took the environment into account when they made decisions about spending our tax money or their profits?

These institutions buy the bulk of America's merchandise and services. So if *they* insisted on green products, companies would have an incentive to make them. And once they were in production, the merchandise would show up in retail stores at a good price, changing the entire economy. Everyone would win.

Well, this process has begun to happen. It's called "Environmentally Preferable Purchasing" (EPP), or "Green Purchasing," and it's actually required by a number of our public and private institutions. It's an encouraging development, but frankly, EPP programs are still barely more than pilot programs scattered around the country. They need a push from the public.

DID YOU KNOW

• Institutions in the U.S. spend enormous sums on goods and services, from office equipment to electricity. In 2002, federal agencies (excluding the military) spent about $350 billion; state and local offices spent over $400 billion. And the nearly 4,000 U.S. colleges and universities added $250 billion in purchasing power.

• It works: In 1993, President Clinton issued the first "green executive order," setting out EPP guidelines for federal agencies to buy recycled office paper, energy-efficient computers and lighting, etc. His standard for paper was 30% recycled fiber. Soon after, 30% recycled paper became readily available to American consumers.

• While EPP programs in federal agencies stalled under the Bush administration, states and municipalities are just getting started. Over the past three years, for example, King County, Washington,

Even the Defense Department's doing it. The "greening" of the New Attack...

(which includes Seattle) spent nearly $55 million in Green Purchasing, saving $1.5 million over conventional products. As King County Executive Ron Sims said, "That's just good government."

• The private sector is getting in on the act. It's no surprise that Ben & Jerry's uses chlorine-free paper for its "eco-pint" containers, but McDonald's has also drastically reduced its packaging and spends over $350 million a year on recycled paper. Toyota and Anheuser-Busch all have green purchasing programs, too.

WHAT YOU CAN DO

Your Partner: The Center for a New American Dream's Responsible Purchasing Network (RPN). Visit them: *responsiblepurchasing.org*

Your Goal: Promote the widespread adoption of EPP. The Responsible Purchasing Network helps institutions develop and implement EPP programs with purchasing guides, resources, and training.

START SIMPLE

A mini-EPP. New American Dream's "6 simple things to get your office to buy": 1) 100% postconsumer recycled, chlorine-free copy paper. 2) Compact fluorescent lightbulbs (at least a few to start, for places where it's harder to change bulbs—because CFLs last so much longer). 3) Unbleached, recycled paper towels and napkins for the lunchroom and bathrooms. 4) Non-toxic cleaning products. 5) EPEAT and ENERGY STAR-rated computers and office electronics. 6) A water filter on the tap to replace bottled water and coolers. For more tips on EPPs, see *50simplethings.com/EPP*

STEPS FOR SUCCESS

Step 1. Do a mini-survey of products and services used where you work. Offer your analysis to the person in charge of purchasing, along with info on products. Go to *50simplethings.com/EPP* for surveys, product guides, and help on approaching decision-makers.

Step 2. EPP your city. Bring RPN guides to a city council meeting, or give the link to your city purchasing agent or mayor's office. Ask for a meeting. RPN can put them in contact with other municipalities that have saved money with EPP programs.

Step 3. Green your university. College campuses have great EPP programs. RPN can help your administrators get one started. Contact *responsiblepurchasing.org*. For K–12: *50simplethings.com/EPP*

...Submarine Program included 80% less solvents and cleaners.

39. OVER THE TOP

At least 1,200 miles of streams in Appalachia have been severely polluted, and 724 miles of streams there have been totally buried *under the millions of tons of rubble created by "mountaintop removal" mining.*

BACKGROUND. How valuable is watching an extra half-hour of television every night? Is it worth sacrificing a mountain?

As you read this, mining companies are blowing the tops off the Appalachian Mountains to get coal...to feed the power plants...to supply the electricity we use to power our TVs, appliances, and lights.

It sounds unbelievable...but it's been going strong since the mid-'80s. Coal companies have been given free rein to turn one of the most beautiful, biodiverse ecosystems in the world into "biologically barren moonscapes." Even parts of the Clean Water Act have been twisted to make "mountaintop removal" easier. The only thing that can stop the coal companies is a loud, sustained public outcry.

DID YOU KNOW

• Mountaintop removal is pretty much just what it sounds like: the tops of mountains are demolished using explosives, so coal-mining companies can reach coal seams underneath.

• Whole mountains are completely decapitated, and the waste created by the explosions is pushed into valleys, burying headwater streams that are crucial to the survival of wildlife and to providing drinking water for local people.

• By 2012, it's projected that 1.4 million acres of mountains and diverse hardwood forests will have been destroyed by mountaintop removal if it continues at its current rate.

• Mountaintop removal endangers more than the natural environment; it threatens nearby communities as well. Mining waste has polluted streams, and nearby residents have suffered repeated flooding. In 2005, three-year-old Jeremy Davidson was crushed to death in his bed by a half-ton boulder that was dislodged at a mountaintop removal operation in Virginia.

More than 50% of America's electricity comes from burning coal.

- In 2000 in Kentucky, a massive lagoon of liquid coal waste stored at a mountaintop removal operation failed. 306 million gallons of toxic coal sludge inundated streams and rivers. The sludge killed all aquatic life for 75 miles and seriously polluted the land and the drinking water of 27,000 people.

WHAT YOU CAN DO

Your Partner: The Ohio Valley Environmental Coalition (OVEC). Since 1987, they've been "dedicated to preserving and protecting our natural heritage." They're fighting to end mountaintop removal. Get acquainted at *ohvec.org*

Your Goal: Join with OVEC to stop mountaintop removal, move to end U.S. reliance on coal, and curb global warming.

START SIMPLE

- *Find your connection.* "People often don't feel connected to what's happening here," says OVEC's Vivian Stockman, "because they live far away. But the truth is, if they get their electricity from a coal-fired plant, they're already involved." Find out how you're connected. Visit *ilovemountains.org/myconnection*
- *Movie night.* Rent or buy *Burning the Future* or other documentaries. Info: *ohvec.org/links/mountaintop_removal/documentaries.html*

STEPS FOR SUCCESS

Step 1. Join forces with OVEC. The only way to tackle mountaintop removal is to stand together. Visit OVEC at *ohvec.org* and sign up for their action alerts. Find out how you can help.

Step 2. Spread the word. Most Americans don't understand the destructiveness of our dependence on coal—the #1 source of global warming gases in the U.S. Take the time to learn…and to tell your family and friends. Fighting mountaintop removal is also an effective way to mobilize against our coal addiction. For more info on coal, check out *energyjustice.net* and *stopmountaintopremoval.org*

Step 3. Support the Clean Water Act. The Clean Water Protection Act, introduced in the House of Representatives in May 2007, is necessary to end the national disgrace of mountaintop removal coal mining in Appalachia. Visit *ilovemountains.org/action/write_your_rep* to see if your representative is a co-sponsor, and write to ask for their support. Find their e-mail address using the links on the page.

40. THE DEVIL AND THE DEEP BLUE SEA

The United Nations estimates that there are 46,000 pieces of plastic per square mile in the world's oceans. Plastic bags in oceans kill a million seabirds and 100,000 sea mammals a year.

BACKGROUND. Most ocean pollution seems complicated. There are oil spills, overfishing, and sewage discharge. There are persistent pollutants that have to be eliminated, like the toxic chemicals that come from smokestacks.

But there's an equally important threat to our oceans that sounds so ordinary, it may be hard to believe. It's garbage. All the bags, bottles, beach toys, food wrappers, and cigarette butts forgotten on the sand, or dropped into a bay, or just left on the ground and washed down a storm drain end up in our oceans. The worst offenders? Plastic bags. According to Greenpeace, Americans discard 100 billion plastic bags a year—only 2% of which are recycled. The rest end up in landfills and, all too often, in our oceans. The solution seems uncomplicated—if we use fewer plastic bags, fewer will get into the oceans.

DID YOU KNOW

• Though oceans cover over 70% of the world's surface, 80% of the pollution in them—including garbage—originates on land.

• The North Pacific Garbage Patch near Hawaii is a vortex made up of 100 million tons of trash—90% of which is plastic—that is drawn by winds and currents from land sources like Japan, Taiwan, and the U.S. Pacific Northwest. A swirling soup of plastic, junk, and dead sea life, it's now larger than the continental U.S.

• Though plastic bags take 1,000 years to completely degrade in the ocean, sun, wind, and water break them into tiny bits—which sea creatures then mistake for plankton, a main food source. There are now 13 lbs. of plastic for every 2 lbs. of plankton in the vortex.

• To make matters worse, persistent pollutants bond to the tiny pieces of plastic, making them toxic. These toxins end up in fish…which end up on our plates, and in us.

WHAT YOU CAN DO

Your Partner: Greenpeace, an environmental group that's been using peaceful direct action and creative communication to protect the planet since 1971. They've led work that banned incineration at sea, ocean dumping, and nuclear testing.

Your Goal: Reduce the amount of pollution that's contaminating our oceans, poisoning marine life, and covering our beaches.

START SIMPLE

No thanks, I don't need a bag. A challenge: Try to go at least one week without accumulating any new plastic bags. (Ideas: Rinse and reuse produce bags, use newspaper delivery bags for pet waste.) If every shopper took just one less bag per month, hundreds of millions of plastic bags could be eliminated as garbage each year.

STEPS FOR SUCCESS

Step 1. Hit the beaches. Participate in Ocean Conservancy's annual International Coastal Cleanup—the world's largest single-day volunteer effort to improve the health of the ocean and its wildlife. Over six million volunteers have removed over 100 million pounds of marine litter from 170,000 miles of beaches and waterways. For details, check out *oceanconservancy.org/icc*

Step 2. Keep it clean. Help keep beaches and waterways clean all year. Join a local chapter of a coastal group like Surfrider, or a local bay or river group; scuba clubs, too. Adopt a beach and agree to clean it at least three times a year. People of all ages can participate—senior citizens to school groups. Links at *50simplethings/ocean*

Step 3. Phase out plastic bags. Work with local retailers. Whole Foods has stopped stocking plastic grocery bags entirely; let your local supermarket know your concerns. Greenpeace has a fact sheet for you to hand out: *greenpeace.org/oceanthreats*

Step 4. Be an advocate. Work to get your town or city council to ban plastic bags. San Francisco has done it, and several other cities are currently considering similar measures. A tax on plastic bags in Ireland led to a 90% drop in their use. Get help at *greenpeace.org/oceanthreats* and *50simplethings/ocean*

Step 5. Help Greenpeace advocate for their biggest priority—marine reserves, protected areas of the ocean where all destructive activities are prohibited. Visit *greenpeace.org/reserve*

41. "IT'S ALIVE!"

Some genetically engineered (GE) plants produce so much of their own pesticide, they're actually classified as pesticides by the EPA.

BACKGROUND. Everyone knows Frankenstein's monster. But most of us are only familiar with it from movies like *Frankenstein vs. Godzilla*, and not the original story. That novel by Mary Shelley, written in 1816, is quite different; it's a cautionary tale about how tinkering with elements of science that we don't understand—and can't control—can lead to disaster.

That's how a lot of people feel about genetically engineered (GE) crops. By manipulating the building blocks of life, genetic scientists have created life forms that could never develop naturally. Genes from bacteria, viruses, plants, animals, and even humans have been engineered into food crops. These creations are being released into the environment and the food supply with little or no regulatory oversight, creating unknown and unpredictable risks. Fortunately, only four major crops—corn, cotton, soy and canola— are widely grown. And it's not too late to take action.

DID YOU KNOW

• Chances are, you've already eaten GE foods. About 70% of processed foods on supermarket shelves contain GE ingredients.

• While Europe, Japan, and others require labels on GE foods, you won't find "genetically engineered" on any label in the U.S. Why? Our government has refused to require labeling.

• GE crops mean more pesticides on our food and in the environment. GE crops can cross-pollinate with related weeds, creating "superweeds" that require more pesticides because they are harder to kill. Since the advent of GE corn, soy, and cotton, annual pesticide use on these crops alone has increased by 122 million pounds.

• The companies that created many GE crops on the market today don't have a stellar safety record. They also produced DDT, PCBs, and Agent Orange. All three had unexpected harmful side effects.

• According to many experts, the biggest problem posed by GE crops today is contaminaton of non-GE and organic crops.

The first field tests of GE foods began in 1992. By 1999, 98.6 million acres...

WHAT YOU CAN DO

Your Partner: The True Food Network, grassroots action network of the Center for Food Safety (CFS), is here to support you with guides for shopping GE-free, taking action in your community, and making your voice heard on important local, state, and national policies. Meet them at *truefoodnow.org* or *centerforfoodsafety.org*

Your Goals: Avoiding GE foods in your own life, joining with your community to protect your local environment from genetic pollution, and demanding government oversight and mandatory labeling at the local, state, and federal level.

START SIMPLE

Empower yourself. If you're like most Americans, you probably don't know very much about GE foods. With help from CFS, you can learn what and where they are. Get CFS's "Pocket Shopper's Guide to Avoiding GE Foods": *truefoodnow.org/shoppersguide*

STEPS FOR SUCCESS

Step 1. Vote with your fork. Fortunately, avoiding many GE foods is simple. Most whole foods—fruits and vegetables, beans, rice, wheat, and other grains—aren't GE. And no organic foods are GE. On the other hand, the majority of nonorganic packaged foods in supermarkets contain GE ingredients: *truefoodnow.org/shoppersguide*

Step 2. Spread the word. The greatest successes often come from simple actions that engage people on a personal level. Talk to friends, family, and co-workers about why GE crops are an important issue to you. You can also gather petition signatures from your friends or even start your own local True Foods group. For downloadable petitions, ideas, and tips: *truefoodnow.org/resources*

Step 3. Take back the aisles. Because of citizen action, supermarkets such as Whole Foods and Trader Joe's have removed GE ingredients from store-brand products. To get involved where you live, download the supermarket activist kit: *truefoodnow.org/supermarkets*

Step 4. Food and politics. Your local, state, and federal government can play an important role in forming policies to regulate and label GE crops and foods. This means your congressional representatives need to hear what you think about genetic engineering. Find your representative, senators, and sample letters at *truefoodnow.org/politics*

...of GE foods had been planted—with no safety testing.

42. HOOK, LINE & SINKER

*In the last 50 years, 90% of the populations of predator fish
like tuna, marlin, cod, halibut, and flounder have disappeared.*

BACKGROUND. Try to imagine the ocean running out of fish. It's impossible, right? And yet, we seem to be headed in that direction. In the last 50 years, about one-third of the fish we catch as seafood have disappeared. Take a moment to think about that.

The main culprit? Our fishing methods. We've been taking too many fish out of the ocean, and not giving them a chance to grow back. The ocean is a natural system, and it *can* regenerate…if we let it. But that means a big change in our approach to fishing.

DID YOU KNOW
• Hundreds of species of fish are in big trouble. Pollution is one reason. Another is global warming, which, in some places, makes water too warm for the animals and plants fish feed on. But the main reason is the unsustainable way we're harvesting them.

• The worst technique is *bottom trawling.* Boats drag a net that can be as wide as a football field, big enough to hold a 747 jet, across the bottom of the ocean. It uproots everything in its path, clearcutting the ocean floor and trapping every marine animal.

• *Gillnetting*, another method of harvesting fish, uses invisible walls of monofilament clear plastic mesh. These nets can be from one to five miles long. They're hung across the ocean; thousands of fish swim into the mesh, get their gills caught, and are trapped—along with turtles and marine mammals.

• One tragic result is the sheer volume of *bycatch*—fish and other marine life (like turtles, dolphins, sea horses) caught and killed needlessly. According to a 2005 study, 25% of all fish pulled from the sea in the United States never make it to market.

• A supposed solution is "fish farms." Farms raising plant-eating fish like tilapia or catfish in ponds can be sustainable. The opposite is true for carnivorous fish raised in huge open-mesh pens in the ocean. One of the worst: salmon farming, which contaminates the ocean floor with huge amounts of fish waste, uses antibiotics

Despite better and better fishing technology, the world's marine catch has fallen since 1988.

heavily, and contributes to overfishing. Tons of "forage fish" are used to feed the salmon—up to 10 pounds of wild fish for each pound of farmed salmon produced.

• If we don't change some of these fishing methods, stop subsidizing fishing fleets, pass better laws and enforce the ones we have, we face a grim future without seafood…and marine life in general.

WHAT YOU CAN DO

Your Partner: Oceana is the largest international ocean environmental advocacy group, dedicated to protecting and restoring the world's oceans. Get acquainted at *oceana.org*

Your Goal: Promote sustainable fishing. "We know how to do it," says an expert. "We just lack will. That's what we have to work on."

START SIMPLE

Eat right. Get a sustainable-seafood cookbook. David Helvarg of Blue Frontier recommends *One Fish, Two Fish, Crawfish, Bluefish* and *The Seafood Lover's Almanac.* Sustainable recipes are also available at *50simplethings.com/overfishing*

STEPS FOR SUCCESS

Step 1. Get a seafood pocket guide from Oceana and Blue Ocean Institute. It tells you which edible fish are depleted, which aren't, and which are contaminated with high levels of mercury. Download it, put it in your wallet, and use it when you shop or go out to eat. Pass it on to friends. Find them at *oceana.org/seafoodguide*

Step 2. Join the sustainable-seafood movement. Work for labeling of healthy and sustainable seafood in grocery stores. Help educate concerned restaurant chefs through a group called Chefs Collaborative. This can have a big impact—about half of the seafood eaten in the U.S. is eaten in restaurants: *chefscollaborative.org*

Step 3. Fight for better policies. The global fishing fleet receives $20 billion in harmful government subsidies annually, allowing it to fish deeper and farther away. Help Oceana stop destructive fishing by joining as an online Wavemaker. Help pressure lawmakers and fishery managers to cut back on subsidies and promote other important ocean issues. Sign up to be a Wavemaker at *oceana.org/north-america/action-center/stay-informed*

100 species of shark are currently threatened with extinction.

43. GOING, GOING...

Try to imagine this: Every day, between 1 and 100
species go extinct somewhere in the world. Every day.

B **ACKGROUND.** Extinction isn't just a scary idea—it's really happening, right now. In fact, experts say we're currently in one of the largest extinction crises in Earth's history. And humans are the cause.

Supporting the Endangered Species Act is one important way to protect species that are teetering on the brink (see p. 34). Here's another: Go out and save some of them. That's right—you can actually find endangered plants or animals in your area, roll up your sleeves, and help them survive.

DID YOU KNOW

• The most authoritative study on extinction, the International Union for the Conservation of Nature, evaluated over 41,000 species. Incredibly, they found 39% of them are now "threatened with extinction."

• This includes a third of our amphibians, a quarter of the world's coniferous trees, an eighth of our birds, and a quarter of all known mammals—even familiar species like the polar bear and hippo.

• Every extinction threatens the web of life. A University of Arizona study found that "with the extinction of a bird, mammal or plant, we aren't necessarily wiping out just one species; we're allowing [many] dependent species to be wiped out as well."

• *Reintroduction* is one of the most successful strategies used to save endangered species. Plants or animals are raised in captivity, then reintroduced to areas from which they've disappeared. It can take years—but it often works. That's what saved the bald eagle.

• A case in point: In Massachusetts, the Plymouth Redbelly Turtle was in trouble; young turtles grown in ponds were being killed by predators. So federal and state wildlife experts set up a program to capture the babies and raise them in captivity. They brought the turtles to schools, where kids raised them, then returned the turtles to their ponds. It was a huge success.

• Another successful technique is simply keeping watch. In Arizona, bald eagle nests are watched by people who camp out during the winter, making sure no one messes with them and that a biologist is notified if an eaglet falls from a nest (so they can put it back).

WHAT YOU CAN DO

Your Partner: The Center for Biological Diversity, the "lean, green endangered species machine," one of North America's premier endangered species advocates. Meet them at *biologicaldiversity.org*

Your Goal: Find endangered species in your area and learn how to protect them. The Center will help.

START SIMPLE

Be a billboard. Most states offer an endangered species or wildlife conservation license plate, with money going to protect local species. Get one: It's a mini-billboard seen by hundreds of people every day.

STEPS FOR SUCCESS

Step 1. Learn more about endangered species. Extinction isn't happening "out there"; it's happening in *your* community. Find out which animals are threatened where you live from the Fish and Wildlife Service. With a little research, you can also find out what's threatening them and who's trying to conserve them. Log on to: *fws.gov/endangered/wildlife.html*

Step 2. Volunteer on a conservation campaign. From a weekend pulling invasive weeds in the Arizona desert to annual campaigns like the Bald Eagle Nest Watch, these efforts are fun, and they really help. Check out your opportunities: *biologicaldiversity.org*

Step 3. Get involved in a reintroduction program. These programs are effective only when coordinated with a combination of expert groups: fish and wildlife services, wildlife refuges, or other partners like zoos or universities. Once a program is designed, it will take time and care…and it might take decades for the species to come back completely, like the Mexican gray wolf or California condor. But it will be worth the effort. For more info: *nfwf.org*

For more resources: *50simplethings.com/extinction*

For a geographically organized list of endangered-species success stories: *esasuccess.org*

44. A NEW THREAT
TO RAINFORESTS

*Deforestation puts more greenhouse gases into the atmosphere
than all of the cars, trucks, planes, trains, and ships on Earth.*

B ACKGROUND. Have you heard of biofuels? They're a
great idea—gasoline substitutes made from plants. And, like
a lot of people, you might think it's good news that big business is finally beginning to produce an alternative fuel.

But in this case, it's not. It turns out that agribusiness is so eager
to "save the Earth" with biofuel that they're tearing down rainforests—which absorb CO_2—to grow the crops that supply it. The
result: Not only are we losing critical forests, but by cutting the
trees down we're creating *more* greenhouse gases than we save by
using biofuels. It's like a bad joke, only it's dead serious.

We need to continue experimenting with biofuels to find a way
to make them environmentally safe. But it's even more pressing
that we get agribusiness out of rainforests right away.

DID YOU KNOW

• Biofuel crops that are having an impact on tropical rainforests
include oil palm, corn, sugarcane, and soy.

• Every ton of palm oil produced results in 33 tons of carbon dioxide emissions—about 10 times more per ton than petroleum. Even
worse, producers are expanding into tropical forests of Indonesia,
Malaysia, and Papua New Guinea at a rate of 2.5 million acres a
year. Since agribusiness began using palm oil for biofuel, palm plantations have become the main cause of deforestation in Indonesia.

• Ethanol, made from corn, is being sold as a "green" fuel. But a
major survey determined that "biofuels produced from food crops
like corn have the least potential of 18 technologies for reducing
carbon emissions over the next 25 years." And the corn required to
make enough ethanol to fill a 25-gallon SUV tank could feed one
person for a year.

• Why do it, then? Profit. Archer Daniels Midland (ADM) lobbied

About 65% of Central America has already been cleared to create pastures for grazing cattle.

the U.S. government for 30 years to promote adding ethanol to gasoline, resulting in $2 billion of government subsidies—most of which has gone directly to ADM.

• As American farmers switch to corn to take advantage of the market for ethanol, they're giving up soy—which is causing a shortage and higher prices. Result: Agribusiness is tearing down rainforests to create new soy fields. Soy has now become a major contributor to deforestation in the Amazon rainforest.

WHAT YOU CAN DO

Your Partner: The Rainforest Action Network (RAN), "some of the most savvy environmental agitators in the business," according to the *Wall Street Journal.* They've been fighting to preserve rainforests since 1985. Get acquainted at *ran.org*

Your Goal: Help spread the word about U.S. agribusiness involvement in rainforest destruction, and pressure them to stop it. RAN's grassroots organizing team is ready to help.

START SIMPLE

Invest in the green. Donate to RAN's Protect-an-Acre Program, a fund that gives small financial grants to indiginous rainforest groups trying to protect their trees: *ran.org/paa*

STEPS FOR SUCCESS

Step 1. Get the facts. Contact RAN to get fact sheets, case studies, a campaign video, and other materials. Join RAN's e-mail list and stay up to speed on the latest campaign activities and action. You can get everything you need at *ran.org/rainforestag*

Step 2. Hold a house party. Invite your friends, neighbors, and family. Show the campaign video, talk about it, and ask people to make a commitment to the rainforest—for example, write a letter to CEOs of the leading U.S. agribusinesses…and ask them to hold their own house parties to expand the campaign. Tips on house parties at *ran.org/rainforestag*

Step 3. Support a moratorium on U.S. incentives for agrofuels. One way to stop this rush to destroy the rainforests is to eliminate government subsidies. To sign an online petition and find out what else you can do, go to *ran.org/agrofuelmoratorium*

As many as 179 tree species can be found in 2.5 acres of rainforest.

45. DO YOUR SHARE(S)

In 2006, as the result of a shareholder resolution filed by a socially responsible investment firm, ExxonMobil was finally forced to admit that fossil fuel consumption is heating up the planet.

BACKGROUND. You might be wondering what shareholder activism has to do with saving the Earth. Well, the truth is, most major polluters are big corporations—and the people they listen to most closely are their shareholders.

Here's the important part: *Anyone* can be a shareholder... including you.

DID YOU KNOW

• Shareholders are actually the owners of any publicly traded company. Legally, all it takes is one share of stock to be considered a shareholder.

• There are limits to what one shareholder can do. That's why, in 1973, a group of religious investors pooled their resources and formed the Interfaith Center for Corporate Responsibility (ICCR). They were the first shareholder group to successfully influence corporations on social issues.

• The ICCR's success inspired eco-activists. After the Exxon *Valdez* oil spill in 1989, the first investor-environmental alliance, the Coalition for Environmentally Responsible Economies (CERES), was formed. Today, fighting for the environment is one of the main areas of shareholder activism.

• Activists use corporate rules to their advantage. Shareholders can file resolutions calling for policy changes or reports—which are then voted on by *all* shareholders. They can also speak during the Q&A session at a company's annual meeting...and sometimes companies listen. For example, Michael Passoff of As You Sow filed a resolution and spoke at a Starbucks shareholders' meeting about their use of milk containing bovine growth hormones. The result: Starbucks researched the alternatives...and ultimately switched to hormone-free milk.

Socially responsible investing is now valued at $2.29 trillion worth of shares.

WHAT YOU CAN DO

Your Partner: As You Sow, founded in 1992 to ensure that corporations and other institutions act "in the long-term best interests of the environment and the human condition." Get to know them at *asyousow.org*

Your Goal: To have an impact on corporate environmental policies. "The challenge," says Michael Passoff, "is to align your investments with your beliefs."

START SIMPLE

Follow the money. Do you own stock? Find out how your investment is affecting the environment. Google the name of a company you've invested in, along with an eco-issue you care about. (If you don't have stock, look up your bank—because after all, that's where your money is.) See what comes up, but don't be surprised if you find good *and* bad news; there are no perfect companies.

STEPS FOR SUCCESS

Step 1. Learn the ropes. Get familiar with things like proxy voting, dialoguing, resolutions, and divestment. How? Start with the two excellent beginner's handbooks at *50simplethings.com/invest*

Step 2. Buy! Buy! You've got to own at least one share of stock to participate in an annual meeting and $2,000 worth of stock to file a resolution. Choose a company on your own, pool money with others (friends, neighbors, co-workers), or join a network of shareholders. However you do it, be sure to vote on shareholder resolutions through proxy ballots when you get the chance. Visit *asyousow.org* for more details.

Step 3. Get in touch. When you're ready, it's time to start speaking up. Writing or calling to discuss your concerns is the first part of what the pros call "dialoguing" with a company. Most corporations have some kind of investor contact. You're the owner—you have a right to contact your company. Learn to "dialogue": *asyousow.org*

Step 4. Show up. Feeling adventurous? Go to a shareholder meeting. Observe or comment, or set up calls or face-to-face meetings with company reps. Find out more at *asyousow.org*

Current estimate: One of every 10 investment dollars is a "socially responsible" investment.

46. WELCOME
TO SOLAR CITY

In 2001, San Francisco voted to put solar on city-owned buildings. Soon after, the city installed a huge solar system on the Moscone Convention Center, and has been installing solar on schools, libraries, and even wastewater treatment plants throughout S.F. ever since.

BACKGROUND. Want to make sure your state and local taxes are spent wisely? Want clean air, local jobs, and lower electricity bills? Here's one way to do it: Get your mayor and city council to switch to solar power. Put solar panels on city hall.

Don't worry—it's not as hard as it sounds. And it's a good economic move; it will pay itself back in 10 to 20 years, then turn a profit for a long time. What a great way to invest in the American solar industry and guarantee our children a clean energy future.

DID YOU KNOW

• Each year, state and local governments spend approximately $12 billion on energy bills. (For counties, it's the second-largest budget item, after salaries.) About 70% of that comes from fossil fuels. So a municipal solar system can save money *and* the environment.

• It would also give the solar industry something it really needs: big orders for equipment...which will enable them to reduce their costs—and their prices. Experience has shown that every time demand for solar power doubles, costs drop 20%. And high prices are the biggest deterrent to solar energy being installed everywhere.

• City governments are perfect for this effort: They're big energy users, they have access to money (they can issue bonds), and they don't mind long-term paybacks. Unlike many businesses, cities can afford to wait for 10 years to make back their initial investment; they know they're still going to be there.

• The most important factor, however, is that doing the right thing is—or should be—one of the core principles of government. Cities should be leading the clean energy revolution by example.

• Grassroots groups have been working with local governments to

In 1921, Albert Einstein won the Nobel Prize for theories related to solar power.

install solar power across the country. One of the most inspiring efforts is in Oroville, a Northern California town with more solar power per capita than any other community in the U.S. They've put solar on their fire dept., police dept., city hall, public works yard, even the Pioneer Museum. If they can do it, so can your town.

WHAT YOU CAN DO

Your Partner: Vote Solar works to bring solar energy into the mainstream by getting municipal governments to invest in it. They're ready to advise and assist anyone who's willing to take on the same challenge. Visit them at *votesolar.org*

Your Goal: Convince politicians in your town or city to join the fight for municipal solar power.

START SIMPLE

Learn to think like a city bureaucrat. How does it feel to have to choose between buying electricity from a utility and going though the hassle of finding funds to invest in a solar system? Familiarize yourself with the basic information you'd need to make that kind of decision. Get a general working knowledge of solar power options. What are the incentives in your state? *50simplethings.com/votesolar*

STEPS FOR SUCCESS

Step 1. Find a champion in city government. Get to know the officials in your city or town. Find one who will make this project their own. Do a little research: What was their campaign platform? What are their interests? Every politician comes into office wondering how to get reelected, or reappointed. This is the project that's going to make them a hero.

Step 2. Put together a coalition of organizations to promote solar. Start with environmental, church, or school groups. You need a broad coalition of voters behind you (or with you) when you meet with your champion.

Step 3. Make an appointment. If you think he or she is approachable, just call for a meeting. If you need to politicize it more, bring a parade of voters along to prove that "solar is super-popular here." From there, it's a matter of keeping track, and having your group keep the pressure on. Vote Solar will help guide you through it.

Two billion people in the world have no access to electricity.

47. WHEN, WHERE, AND HOW MANY?

In more than half the developing countries studied by the U.N., population has been growing faster than food supplies.

BACKGROUND. Population growth is a controversial subject, but it's really impossible to discuss saving the Earth without addressing it. There's an obvious connecton between the growing number of people on Earth and problems with our global life-support system: More people means more resources depleted, more pollution created, more habitat destroyed.

Many thoughtful Americans believe that human population growth (now at around 78 million a year) is the most important environmental issue facing our world; until we deal with it, they say, nothing else we do will matter.

DID YOU KNOW

• It took almost all of human history—until the early 1800s—for the world's population to reach 1 billion. Then it exploded: By 1960 there were 3 billion people. By 1990 there were 6 billion. And the United Nations predicts we'll hit 9 billion by mid-century.

• Rapid population growth puts enormous pressure on essential resources. Humans are already using about half the accessible freshwater available in rivers, lakes, and aquifers. It's estimated that by 2025, it will be at least 70%; by 2050 it could be 90%.

• According to a recent study, population growth has *already* resulted in the global loss of an area of prime farmland equal to about two-thirds of North America.

• In less-developed areas, where about 80% of the global population now lives, forests are disappearing rapidly. Nearly half of our planet's forested area is now gone.

• How do we stabilize human population? Experience shows that when people are given access to family planning information and services, average family size drops. Singapore went from 6.4 children per family in 1950 to 1.35 in 2005. Thailand went from 6.4 to 1.83 in the same period. Mexico went from 6.97 to 2.4.

According to some estimates, population growth alone will account for 35% of...

WHAT YOU CAN DO

Your Partner: Population Connection believes humanity's well-being, and even survival, depend on finding an equilibrium between population and the environment. Get to know them: *popconnect.org*

Your Goal: Make population information and services available to anyone who wants them. Help others to make the connection between the health of our environment and world population.

START SIMPLE

• *Arm yourself with the facts.* Don't be intimidated; be an authority. Start learning at *popconnect.org* and get more resources at *50simplething.com/population*

• *Knowledge is power.* Share your thoughts with your members of Congress. There is no substitute for a real letter from a real voter. Become known as the "Population Person." For more information, visit *capwiz.com/zpg/dbq/officials*

STEPS FOR SUCCESS

Step 1. Think locally. 1) When you go to a drugstore, ask the pharmacist or store manager if they fill presciptions for birth control (and emergency contraception). If they don't, change pharmacies, and tell them why. 2) Oppose abstinence-only sex education in schools. Experts say it doesn't prevent pregnancies. In fact, it may have the opposite effect. Get info at *SIECUS.org*

Step 2. Think nationally. The U.S., with 5% of the global population, uses about 25% of the world's fossil fuel resources, and the average American consumes 120 lbs. a day in natural resources. So population stabilization in the U.S. could have a big impact on protecting the environment. Population Connection recommends an increase in federal funding for family planning, and that all women and couples make their own reproductive decisions. Policy recommendations at their Legislative Action Center: *capwiz.com/zpg/home*

Step 3. Think globally. The U.S. should be a leader in providing foreign aid to promote population stabilization. Instead, President Bush imposed the Global Gag Rule, which prevents international family planning groups from providing women with comprehensive medical information. Read fact sheets on related topics. Go to *popconnect.org* and click on "fact sheets."

...the global increase in CO_2 emissions between 1985 and 2100.

48. TURN UP
THE HEAT

The glaciers on Mount Kilamanjaro, in Africa, are melting so quickly that the mountain lost a quarter of its ice from 2000 to 2006. In the Andes Mountains, glaciers are melting 10 times faster than they did just 20 years ago.

BACKGROUND. Are politicians really paying enough attention to global warming? From where *we* sit, it seems like the most serious environmental issue of our time...yet there's been little action on it. How much glacier ice has to melt before Washington starts taking it seriously?

Maybe it's time for a new strategy. Maybe the time has come for us to stand up, raise our voices, and tell the politicians what we really want—take a page from historic social movements like women's suffrage and Civil Rights and gather in public to demand an end to the destructive policies that keep us spewing greenhouse gases into the atmosphere. Maybe it's time to turn our passion into a real movement so that with our combined efforts we can create a productive clean-energy economy. If you're ready to hit the streets, there are people waiting to join you.

DID YOU KNOW

• Global warming is becoming a big political issue. According to a *Washington Post* poll in 2007, "7 in 10 Americans want more federal action on global warming, and about half of those surveyed think the government should do 'much more' than it is doing now."

• About a third of the people polled said that global warming is the most important isue we face today—twice as many as the year before. Other polls show similar results. Almost all show a huge increase in support for action.

• Public demonstrations have always been an effective way to galvanize latent support for an issue. From the Boston Tea Party to Earth Day 2000, mass rallies and small protests alike have had huge political impact. Now global warming activists are mobilizing, with events like the Mother's Day 2008 action coordinated by 1Sky.

A matter of priority? The 2008 research budget proposed by the Dept. of Energy was $1.5 billion...

WHAT YOU CAN DO

Your Partner: 1Sky, a national campaign to catalyze a society-wide movement in support of federal action to tackle global warming. Learn more at *1sky.org*

Your Goal: Become a global warming activist—organize or participate in public events, and help mobilize others to do the same.

START SIMPLE

Movie night. Two recommended movies on global warming: The Oscar-winning *An Inconvenient Truth*, which features Al Gore, and *Everything's Cool*, "a real-life disaster movie" featuring other prominent global warming educators. Both are available on DVD.

STEPS FOR SUCCESS

Step 1. Join the movement. Sign up for 1Sky's *Activist Newsletter*. Get updates and alerts about the national movement and information on joining local efforts. For info on additional newsletters, go to *50simplethings.com/activist*

Step 2. Find out what's happening where you live. Tap into 1Sky's open-source online calendar, where organizers can list events and community members can find local actions. It's a clearinghouse for global warming actions nationwide. Go to *1sky.org/action* for activist toolkits, talking points, materials, and details on national organizing conference calls.

Step 3. Spread the word. The Climate Project is another nonprofit dedicated to educating the public about this issue. More than 1,000 volunteers have been trained by Al Gore to present a version of the slide show on which *An Inconvenient Truth* is based. Arrange a presentation for your school, community group, or office. Visit *theclimateproject.org* and follow the steps to "request a presentation."

Step 4. Join in on a congressional alert day. 1Sky periodically organizes days when people around the nation simultaneously contact their members of Congress. Put a group together to visit your congressperson's district office and present the 1Sky platform. For a toolkit including talking points and materials, go to *1sky.org/action*

Step 5. Become a Lead Volunteer. 1Sky relies on grassroots climate groups to organize events "on the ground." Become a Lead Volunteer with a local group and help bring out volunteers and attendees to National Action Days. For details, see *1sky.org/allies*

...By that time, we had already spent $450 billion on the war in Iraq.

49. TARGET: ZERO WASTE

*On average, every pound of materials you take to the curb
(as trash or recycling) equals 71 pounds of waste created
during extraction, manufacturing, and production.
That's 70 pounds of waste you never see.*

BACKGROUND. To some people, recycling represents both the strengths and weaknesses of the environmental movement: On one hand, it's a successful grassroots effort that saves resources and reflects people's willingness to make changes for ecological reasons. On the other hand, it's a substitute for making *real* changes in the way we approach the whole concept of waste.

"Recycling is important, but it takes place too far down the waste stream," says Linda Christopher of the GrassRoots Recycling Network. "By the time something can be recycled, we've already used up water and energy, created pollution, and caused deforestation. If we really want to make a change, the important decision points are not after we've brought a product into our homes—it's 'upstream,' where the product is made...and even further out—where it was designed. We need to stop *managing* waste and start *eliminating* it."

This thinking is called "Zero Waste." It doesn't mean that anyone expects us to produce no waste. It's a way of going beyond recycling, formally acknowledging that waste isn't inevitable—that it can be reduced by designing it out of the system in the first place.

DID YOU KNOW

• As of 2005, nearly 70% of U.S. municipal waste was dumped in landfills or burned in incinerators; only 30% was recycled.

• A 70/30 ratio may not sound bad, but Americans create almost 2 tons of waste per person every year. That comes to over 400 million tons of municipal waste—two-thirds of which is thrown away.

• But municipal waste accounts for, at most, 20% of the waste we

By 2007, more than 20 cities had adopted Zero Waste policies.

each generate. Almost 95% of industrial materials become waste before a product is manufactured, and 80% of what we make is thrown away within six months of production.

• Zero Waste is good for the economy: Studies show that 10,000 tons of solid waste can create 4 composting jobs, 10 recycling jobs, and up to 250 reuse jobs...but only one landfill or incinerator job.

WHAT YOU CAN DO

Your Partner: The GrassRoots Recycling Network (GRRN), a leading national nonprofit working toward Zero Waste: *grrn.org*

Your Goal: Reduce the waste in your own life, and help your community move toward a new value of reduced waste.

START SIMPLE

• *Educate yourself.* How is Zero Waste different from recycling? Watch this six-minute video: *ecocycle.org/zerowastevideo*

• *Movie night.* Rent or buy the award-winning documentary *Manufactured Landscapes.* It's a fascinating film that will give you a worldwide perspective on waste.

STEPS FOR SUCCESS

Step 1. Reduce! Find simple ways to cut down on waste. Here are 3: 1) Bring reusable bags to the supermarket (Tip: Keep two sets, so there's always one in your car). 2) Replace throwaway coffee cups with a travel mug. 3) Buy a water filter and a stainless steel reusable bottle instead of expensive plastic-bottled water (see p. 80). For more tips: *50simplethings.com/zerowaste*

Step 2. Hold a Zero Waste event. Demonstrate "Zero Waste in action" by organizing an event so there's no (or minimal) garbage created. Get a Zero Waste event kit and instructions from our partners at Eco-Cycle: *ecocycle.org/zwevents*

Step 3. For the ambitious. Advocate to make your community a Zero Waste city, like San Francisco and Seattle. Join forces with a community group such as a nonprofit recycling center. Ask your city council to pass a resolution which sets a Zero Waste goal. GRRN has all the information you need, and will help train and prepare you: *grrn.org*

Heavy estimate: Americans throw away over 100 million tons of paper each year.

50. GREEN-COLLAR JOBS

"We want to use the green-collar movement to move people out of poverty. Little green fairies do not come out of the sky and install solar panels. Someone has to do the work."
—Majora Carter, Sustainable South Bronx

BACKGROUND. What's the best way to give Americans of all socioeconomic backgrounds a tangible stake in fighting for issues like global warming?

Easy: Make it their livelihood. Every day, about 135 million people go to work in the U.S. Imagine what would happen if millions of those jobs—plus new ones created for people who are currently unemployed—were in fields like renewable energy, sustainable agriculture, and green building. Our two crucial concerns about survival—the environment and making a living—would be combined. A person's commitment to their job would also be their commitment to the planet.

Right now, there's a great opportunity not only to make America's economy stronger by making it greener, but to make Americans living in poverty part of a revitalized middle class. The first thing we have to do is provide the training that will turn 20th-century blue-collar jobs into secure 21st-century green-collar jobs.

DID YOU KNOW

• There's already a huge green economy developing. In 2006 renewable energy and energy efficiency technologies generated 8.5 million new jobs, nearly $970 billion in revenue, and more than $100 billion in industry profits.

• According to the National Renewable Energy Lab, the major barriers to a more rapid adoption of renewable energy and energy efficiency in America are insufficient skills and training.

• We've already taken a first step to rectify that. In December 2007, President Bush signed the Green Jobs Act to train workers for green-collar jobs. It authorizes $125 million for workforce training programs targeted to veterans, displaced workers, at-risk youth, and families in extreme poverty. It will train people for jobs like installing solar panels and weatherization.

What are green-collar jobs? Bicycle repair, energy retrofits, furniture-making using environmentally...

• Sustainable South Bronx, an environmental group founded in 2001 by Majora Carter, is one of a number of groups that are making green-collar jobs real. It has trained 70 former drug addicts, welfare recipients, and convicts for jobs in ecological restoration, green roof installation, and hazardous waste cleanup.

WHAT YOU CAN DO

Your Partner: Green for All. Their goal is to build a green economy strong enough to lift people out of poverty. They support job training, employment, and entrepreneurial opportunities in the emerging green economy for everyone.

Your Goal: Help create an equitable green economy.

START SIMPLE

• **Movie night.** Watch the *Everything's Cool* DVD. This documentary about addressing global warming includes a special extra on the Green Jobs Revolution. Details: *everythingscool.org/article.php?id=38*.

• **Watch Van Jones**, founder of Green for All, speak on EcoEquity: *youtube.com/watch?v=2SmF3B3734E*

STEPS FOR SUCCESS

Step 1. Get involved. This is a movement in its infancy, so it needs all of us to get the word out. Sign up with Green for All and its partners, like the Apollo Alliance. If you're a college student (or just college-age) join the Energy Action Coalition. They'll keep you up to date on news, events, and what you can do. Find them at *greenforall.org/resources/whatcanido.html* or *50simplethings.com/greencollar*

Step 2. Make it legal. Green for All's Action Center tracks the status of pending legislation on green-collar jobs. Voice your support for full funding of green-collar jobs creation and training through Letters to the Editor and e-mails to elected officials. You can find sample letters and contact info at *greenforall.org/resources/*

Step 3. The Big Goal. Green for All and its partners are proposing a national Clean Energy Corps (CEC) Initiative, which would be funded by the federal government at $20 billion a year for 10 years. You can help build a CEC from the grassroots up by supporting efforts to set up job training programs for people to learn green-collar skills like installing solar panels, retrofitting houses, and more. More info at *50simplethings.com/greencollar* and *greenforall.org*

...certified and recycled wood, parks and open space maintenance, printing with non-toxic inks, etc.

MEET YOUR PARTNERS

We asked the groups that joined us in creating this book to introduce themselves to you. Here they are, in alphabetical order.

1SKY is dedicated to magnifying the climate change movement into a massive society-wide mobilization by communicating a positive vision and passing a coherent set of national policies by 2009 that rise to the scale of the challenge. To achieve such change, we must mobilize new participants to join the movement, including great representation from communities of faith, social justice, human rights, business, artists, students, labor, and health. The policies we promote will create 5 million green jobs that provide pathways out of poverty, reduce carbon emissions by at least 25% by 2020 and at least 80% by 2050, and end the creation of new dirty coal plants. *1sky.org*

THE ALLIANCE TO SAVE ENERGY is a nonprofit coalition of prominent business, government, environmental, and consumer leaders who promote the efficient and clean use of energy worldwide to benefit consumers, the environment, the economy, and national security. The Alliance emphasizes that energy efficiency is the world's most critical energy resource and that energy efficiency is the quickest, cheapest, cleanest way to extend our world's energy supplies. A strength of the Alliance is the ability to create partnerships and to bring various constituencies and parties together to advance the overall goal of moving the United States and the world toward a more energy-efficient, sustainable future. *ase.org*

AMERICAN RIVERS is the only national organization standing up for healthy rivers so our communities can thrive. Through national advocacy, innovative solutions, and our growing network of strategic partners, we protect and promote our rivers as valuable assets that are vital to our health, safety, and quality of life. Founded in 1973, American Rivers has more than 65,000 members and supporters nationwide. *americanrivers.org*

AMERICAN SOLAR ENERGY SOCIETY is leading the renewable energy revolution. ASES is a nonprofit organization dedicated to increasing the use of solar energy, energy efficiency, and other sustainable technologies in the U.S. Learn more at *ases.org*

AS YOU SOW is a leading practitioner of shareholder advocacy for social and environmental resolutions and has engaged top investors and senior management at dozens of companies such as Apple, Time Warner, Starbucks, and Home Depot, among others. As You Sow mobilizes investor coalitions, leads dialogues with company executives, files shareholder resolutions, conducts large-scale shareholder solicitation campaigns, develops educational materials, and conducts media outreach to promote more responsible corporate behavior. *asyousow.org*

AUDUBON. Now in its second century, Audubon connects people with birds, nature, and the environment that supports us all. Our national network of community-based nature centers, chapters, scientific, education, and advocacy programs engages millions of people from all walks of life in conservation action to protect and restore the natural world. *audubon.org*

THE CENTER FOR BIOLOGICAL DIVERSITY believes that the welfare of human beings is deeply linked to nature—to a vast diversity of wild animals and plants. We work to secure a future for all species, great and small, hovering on the brink of extinction. We do so through science, law, and creative media, with a focus on protecting the lands, waters, and climate that species need to survive. *biologicaldiversity.org*

CENTER FOR HEALTH, ENVIRONMENT & JUSTICE mentors a movement building healthier communities by empowering people to prevent harm for as many people as possible. CHEJ empowers groups by providing the tools, direction, and encouragement they need to promote human health and environmental integrity. Following her successful effort to prevent further harm for people living in contaminated Love Canal, Lois Gibbs founded CHEJ to continue the journey. CHEJ has assisted over 10,000 groups nationwide. *chej.org*

THE CENTER FOR A NEW AMERICAN DREAM helps individuals and institutions consume responsibly for a better world. Live consciously, buy wisely, and make a difference. Improving the quality of your life by making smarter choices as a consumer has never been more important. And we all need to remember that every dollar spent (or not) is a vote for the environment and every purchase has an impact on our planet. For tips on getting more of what matters, visit us at *newdream.org*

CO-OP AMERICA has been pioneering the green economy since 1982—helping millions of people live greener lives, stopping corporate abuse, building fair trading systems, ending sweatshops, fighting climate change, and doing much, much more to create a world that is both socially just and environmentally sustainable. *coopamerica.org*

DEFENDERS OF WILDLIFE's mission is to protect all native species and the habitats upon which they depend. With more than one million members and activists, Defenders of Wildlife is a leading advocate for innovative solutions to safeguard our wildlife heritage for generations to come. For more information, visit *defenders.org*

EARTH ISLAND INSTITUTE's International Marine Mammal Project works to protect whales, dolphins, and their marine habitat. Earth Island established the global Dolphin Safe tuna program that ensures dolphins are not harassed, netted, or killed in tuna nets, and led the efforts to return Keiko, the orca star of the hit movie *Free Willy*, to his home waters off Iceland. Earth Island is a leader in the fight to end whaling and killing of dolphins. *earthisland.org*

EARTHJUSTICE is the nation's leading nonprofit environmental law firm. We represent hundreds of communities and organizations in court in cases that protect our air, water, public lands, and wildlife. Visit us at *earthjustice.org*

EARTHWORKS is a nonprofit organization dedicated to protecting communities and the environment from the destructive impacts of mineral development, in the U.S. and worldwide. We fulfill our mission by working with communities and grassroots groups to reform government policies, improve corporate practices, influence investment decisions, and encourage responsible materials sourcing and consumption. *earthworksaction.org*

ECO-CYCLE works to identify, explore, and demonstrate sustainable resource management through the concepts and practices of Zero Waste. We are changing the rules of our society to eliminate wasteful and polluting practices and building infrastructure for reuse, recycling, and composting to replace landfills and incinerators. Eco-Cycle is partnering with local businesses, municipalities, schools, and individuals to create a Zero Waste community model for the rest of the world. *ecocycle.org*

THE ENDANGERED SPECIES COALITION is a national network of hundreds of conservation, scientific, religious, sporting, business, and community organizations working to protect endangered species and habitats. Through public education, scientific information, and citizen participation, we work to protect our nation's wildlife, fish, and plants on the brink of extinction. *stopextinction.org*

ENVIRONMENTAL CONCERN promotes public understanding and stewardship of wetlands with the goal of improving water quality and enhancing nature's habitat through wetland outreach and education, native species horticulture, and the restoration, construction, and enhancement of wetlands. *wetland.org*

THE ENVIRONMENTAL LAW & POLICY CENTER is the Midwest's leading environmental, legal advocacy, and economic development organization. *elpc.org*

THE ENVIRONMENTAL WORKING GROUP uses the power of information to protect human health and the environment. EWG's three programs—Toxics & Health, Sustainable Agriculture, and Natural Resources—aim to reform chemical policies, shift government subsidies, and protect your public land. We also educate consumers online and through our e-mail bulletin about what you can do to limit chemical exposure. See it all and sign up at *ewg.org*

FOOD & WATER WATCH is a nonprofit consumer organization that works to ensure clean water and safe food. We challenge the corporate control and abuse of our food and water resources by empowering people to take action and by transforming the public consciousness about what we eat and drink. *foodandwaterwatch.org*

FOODROUTES CONSERVANCY. Where does your food come from? Join the growing number of active volunteers at FoodRoutes Conservancy working to support, conserve, and restore our nation's local food systems. Buy fresh, buy local, and support a sustainable food system that will feed us for generations to come. *foodroutes.org*

FRIENDS OF THE EARTH is the U.S. voice of an influential, international network of grassroots groups in 70 countries. Founded in San Francisco in 1969 by David Brower, Friends of the Earth has for decades been at the forefront of high-profile efforts to create a more healthy, just world. Our members were the founders of what is now the world's largest federation of democratically elected environmental groups, Friends of the Earth International. *foe.org*

THE GRASSROOTS RECYCLING NETWORK is a leading national organization in North America working to implement Zero Waste. GRRN has resources for Zero Waste Communities, Zero Waste Businesses, and Zero Waste Campuses. *grrn.org*

GREEN FOR ALL has a simple but ambitious mission: to help build a green economy strong enough to lift people out of poverty. By advocating for a national commitment to job training, employment, and entrepreneurial opportunities in the emerging green economy—especially for people from disadvantaged communities—we fight both poverty and pollution at the same time. *greenforall.org*

GREENPEACE is an independent campaigning organization with 2.7 million members worldwide that uses peaceful direct action and creative communication to expose global environmental problems and promote solutions for the future. *greenpeace.org/usa*

THE INSTITUTE FOR LOCAL SELF-RELIANCE. Since 1974, ILSR has been working to enable communities with tools to increase economic effectiveness, reduce waste, decrease environmental impact, and provide for local ownership of the infrastructure and resources essential for community well-being. ILSR has offices in Washington, D.C., and Minneapolis, MN. *ilsr.org*

INTERFAITH POWER AND LIGHT. The mission of IPL is to be faithful stewards of Creation by responding to global warming through the promotion of energy conservation, energy efficiency, and renewable energy. This ministry intends to protect the Earth's ecosystems, safeguard public health, and ensure sufficient, sustainable energy for all. *interfaithpowerandlight.org*

THE IZAAK WALTON LEAGUE. Through a network of nearly 300 community-based chapters, and with support from national and regional staff, IWLA works on a broad range of conservation issues through education, advocacy, and on-the-ground projects for volunteers. On everything from water quality to wilderness protection to renewable energy, the League has been a leading voice in conservation for more than eight decades. *iwla.org*

THE LEAGUE OF CONSERVATION VOTERS is turning environmental values into national priorities. To secure the environmental future of our planet, LCV's mission is to advocate for sound environmental policies and to elect pro-environmental candidates who will adopt and implement such policies. *lcv.org*

THE NATIONAL CAMPAIGN FOR SUSTAINABLE AGRICULTURE brings people's voices to policy development for sustainable agriculture. We strengthen family farms and rural communities building a safe, just, and sustainable food system that connects healthy communities, both rural and urban. *sustainableagriculture.net*

NATURAL RESOURCES DEFENSE COUNCIL is one of the nation's most effective environmental action organizations. We use law, science, and the support of 1.2 million members and online activists to protect the planet's wildlife and wild places, and to ensure a safe and healthy environment for all living things. *nrdc.org*

NATIONAL WILDLIFE FEDERATION has over 4 million members, partners, and supporters who actively educate, inspire, and promote achievable solutions to everyday Americans in communities from coast-to-coast. Our conservation work focuses on three major areas that will have the biggest impact on the future of America's wildlife: confronting global warming, protecting and restoring wildlife habitats, and connecting people with nature. *nwf.org*

OCEANA campaigns to protect and restore the world's oceans. Our team of marine scientists, economists, lawyers, and advocates win specific and concrete policy changes to reduce pollution and to prevent the irreversible collapse of fish populations, marine mammals, and other sea life. Visit us at *oceana.org*

OHIO (RIVER) VALLEY ENVIRONMENTAL COALITION members have a vision for our future, a commitment to greater democracy, and a cleaner environment. We are people from all walks of life who take an active role in community by educating ourselves and others about the effects of pollution and corrupt politics on the environment and human health and by taking unified action on regional environmental threats. *ohvec.org*

PESTICIDE ACTION NETWORK works to replace hazardous pesticides with ecologically sound and socially just alternatives. As one of five PAN Regional Centers worldwide, PAN North America links local and international consumer, labor, health, environment, and agriculture groups into an international citizens' action network. *panna.org*

POPULATION CONNECTION. Overpopulation threatens the quality of life for people everywhere. Population Connection is the national grassroots population organization that educates young people and advocates progressive action to stabilize world population at a level that can be sustained by Earth's resources. *populationconnection.org*

RAINFOREST ACTION NETWORK runs hard-hitting campaigns to break America's oil addiction, reduce our reliance on coal, protect endangered forests and Indigenous rights, and stop destructive investments around the world through education, grassroots organizing, and nonviolent direct action. *ran.org*

THE RAINFOREST ALLIANCE works to conserve biodiversity and ensure sustainable livelihoods by transforming land-use practices, business practices, and consumer behavior. *rainforest-alliance.org*

SEACOLOGY protects island coral reefs and rainforests by providing a tangible benefit that island villagers request, such as an elementary school, health clinic, or fresh water delivery system, in exchange for the establishment of a marine or forest reserve. Seacology has launched over 180 such win-win projects preserving two million acres of threatened coral reef and rainforest habitat, thus both protecting fragile ecosystems and improving the quality of life of islanders. This is why noted marine biologist Dr. John McCosker has stated that "dollar for dollar, pound for pound, Seacology gets more output than any conservation group that I've seen." *seacology.org*

THE SIERRA CLUB is the nation's oldest and largest grassroots environmental organization. Our members and supporters are more than 1.3 million of your friends and neighbors. With over 800 Cool Cities and counting, we are solving global warming one city at a time with innovative and money-saving ideas. *sierraclub.org*

TRANSFAIR USA enables sustainable development and community empowerment by cultivating a more equitable global trade model that benefits farmers, workers, consumers, industry, and the Earth. We achieve our mission by certifying and promoting Fair Trade products. *transfairusa.org*

TREEPEOPLE. Founded in 1973 by teenagers, TreePeople gives communities the tools to take action in their backyards, schools, parks, and neighborhoods. TreePeople has planted more than two million trees in the Los Angeles area and inspires children, teens, and adults with urban forestry, environmental education, and sustainability programs. Simply put, their work is about helping nature heal our cities. *treepeople.org*

THE TRUE FOOD NETWORK is the grassroots action arm of the Center for Food Safety, a national nonprofit public interest and environmental organization focused on challenging harmful food production technologies and promoting sustainable alternatives through legal actions, policy analysis, and grassroots citizen action. *truefoodnow.org* and *centerforfoodsafety.org*

THE TRUST FOR PUBLIC LAND is a national nonprofit land conservation organization that conserves land for people to enjoy as parks, community gardens, historic sites, rural lands, and other natural places, ensuring livable communities for generations to come. Since 1972, TPL has worked with willing landowners, community groups, and national, state, and local agencies to complete more than 3,500 land conservation projects in 47 states, protecting more than two million acres. Since 1994, TPL has helped states and communities craft and pass over 330 ballot measures, generating more than $25 billion in new conservation-related funding. *tpl.org*

THE U.S. GREEN BUILDING COUNCIL. The vision of USGBC is a sustainable built environment within a generation. Its membership includes corporations, builders, universities, government agencies, and other organizations. Since USGBC's founding in 1993, the Council has grown to more than 13,000 member companies and organizations, a comprehensive family of LEED® green building rating systems, an expansive educational offering, the industry's popular Greenbuild International Conference and Expo, and a network of 72 local chapters and affiliates. *usgbc.org*

THE UNION OF CONCERNED SCIENTISTS (UCS) is a nonprofit partnership of scientists and citizens combining rigorous scientific analysis, innovative policy development, and effective citizen advocacy to achieve practical environmental solutions. UCS works to ensure that all people have clean air, energy, and transportation, as well as food that is produced in a safe and sustainable manner. We strive for a future that is free from the threats of global warming and nuclear war, and a planet that supports a rich diversity of life. In short, UCS seeks a great change in humanity's stewardship of the Earth. *ucsusa.org*

VOTE SOLAR's mission is to bring solar energy into the mainstream. Think of America's problems. Global warming. Air pollution. Fossil fuel addiction. Think about how many could be helped by a transition to clean, renewable solar energy. The clock is ticking. *votesolar.org*

WATERKEEPER ALLIANCE. Everyone has the right to clean water. Waterkeeper Alliance acts locally and organizes globally to fight for clean water and strong communities. *waterkeeper.org*

WILDEARTH GUARDIANS protects and restores the wildlife, wild places, and wild rivers of the American West. *wildearthguardians.com*

THE WILDERNESS SOCIETY. The mission of TWS is to protect wilderness and inspire Americans to care for our wild places. Since 1935, TWS has helped protect more than 105 million acres for future generations. TWS has regional offices all around the country; find contact information for the office near you at *wilderness.org*

Visit our Web site at

50simplethings.com

You're not alone...

We'll help you find a
way to stay involved,
and stay effective.

Together, we can
make a difference.